ETHICAL AND LEGAL CONSIDERATIONS IN MITIGATING PANDEMIC DISEASE

Workshop Summary

Stanley M. Lemon, Margaret A. Hamburg, P. Frederick Sparling, Eileen R. Choffnes, and Alison Mack, *Rapporteurs*

Forum on Microbial Threats
Board on Global Health

INSTITUTE OF MEDICINE
OF THE NATIONAL ACADEMIES

THE NATIONAL ACADEMIES PRESS
Washington, D.C.
www.nap.edu

THE NATIONAL ACADEMIES PRESS 500 Fifth Street, N.W. Washington, DC 20001

NOTICE: The project that is the subject of this report was approved by the Governing Board of the National Research Council, whose members are drawn from the councils of the National Academy of Sciences, the National Academy of Engineering, and the Institute of Medicine.

This project was supported by contracts between the National Academy of Sciences and the U.S. Department of Health and Human Services: National Institutes of Health, National Institute of Allergy and Infectious Diseases, Centers for Disease Control and Prevention, and Food and Drug Administration; U.S. Department of Defense: Global Emerging Infections Surveillance and Response System, Walter Reed Army Institute of Research, and Defense Threat Reduction Agency; U.S. Department of Veterans Affairs; U.S. Department of Homeland Security; U.S. Department of State; Lawrence Livermore National Laboratory; American Society for Microbiology; Sanofi Pasteur; Burroughs Wellcome Fund; Pfizer; GlaxoSmithKline; Infectious Diseases Society of America; and the Merck Company Foundation. Any opinions, findings, conclusions, or recommendations expressed in this publication are those of the author(s) and do not necessarily reflect the view of the organizations or agencies that provided support for this project.

International Standard Book Number-13: 978-0-309-10769-3
International Standard Book Number-10: 0-309-10769-5

Additional copies of this report are available from the National Academies Press, 500 Fifth Street, N.W., Lockbox 285, Washington, DC 20055; (800) 624-6242 or (202) 334-3313 (in the Washington metropolitan area); Internet, http://www.nap.edu.

For more information about the Institute of Medicine, visit the IOM home page at: **www.iom.edu.**

Printed in the United States of America.

The serpent has been a symbol of long life, healing, and knowledge among almost all cultures and religions since the beginning of recorded history. The serpent adopted as a logotype by the Institute of Medicine is a relief carving from ancient Greece, now held by the Staatliche Museen in Berlin.

COVER: A detailed section of a stained glass window 21″ × 56″ depicting the natural history of influenza viruses and zoonotic exchange in the emergence of new strains was used to design the front cover. Based on the work done at St. Jude Children's Research Hospital supported by American Lebanese Syrian Associated Charities (ALSAC) and the National Institute of Allergy and Infectious Diseases (NIAID). Artist: Jenny Hammond, Highgreenleycleugh, Northumberland, England.

Suggested citation: Institute of Medicine (IOM). 2007. *Ethical and Legal Considerations in Mitigating Pandemic Disease, Workshop Summary*. Washington, DC: The National Academies Press.

"Knowing is not enough; we must apply.
Willing is not enough; we must do."
—Goethe

INSTITUTE OF MEDICINE
OF THE NATIONAL ACADEMIES

Advising the Nation. Improving Health.

THE NATIONAL ACADEMIES

Advisers to the Nation on Science, Engineering, and Medicine

The **National Academy of Sciences** is a private, nonprofit, self-perpetuating society of distinguished scholars engaged in scientific and engineering research, dedicated to the furtherance of science and technology and to their use for the general welfare. Upon the authority of the charter granted to it by the Congress in 1863, the Academy has a mandate that requires it to advise the federal government on scientific and technical matters. Dr. Ralph J. Cicerone is president of the National Academy of Sciences.

The **National Academy of Engineering** was established in 1964, under the charter of the National Academy of Sciences, as a parallel organization of outstanding engineers. It is autonomous in its administration and in the selection of its members, sharing with the National Academy of Sciences the responsibility for advising the federal government. The National Academy of Engineering also sponsors engineering programs aimed at meeting national needs, encourages education and research, and recognizes the superior achievements of engineers. Dr. Wm. Wulf is president of the National Academy of Engineering.

The **Institute of Medicine** was established in 1970 by the National Academy of Sciences to secure the services of eminent members of appropriate professions in the examination of policy matters pertaining to the health of the public. The Institute acts under the responsibility given to the National Academy of Sciences by its congressional charter to be an adviser to the federal government and, upon its own initiative, to identify issues of medical care, research, and education. Dr. Harvey V. Fineberg is president of the Institute of Medicine.

The **National Research Council** was organized by the National Academy of Sciences in 1916 to associate the broad community of science and technology with the Academy's purposes of furthering knowledge and advising the federal government. Functioning in accordance with general policies determined by the Academy, the Council has become the principal operating agency of both the National Academy of Sciences and the National Academy of Engineering in providing services to the government, the public, and the scientific and engineering communities. The Council is administered jointly by both Academies and the Institute of Medicine. Dr. Ralph J. Cicerone and Dr. Wm. Wulf are chair and vice chair, respectively, of the National Research Council.

www.national-academies.org

FORUM ON MICROBIAL THREATS

STANLEY M. LEMON (*Chair*), School of Medicine, University of Texas Medical Branch, Galveston

MARGARET A. HAMBURG (*Vice-chair*), Nuclear Threat Initiative/Global Health & Security Initiative, Washington, DC

P. FREDERICK SPARLING (*Vice-chair*), University of North Carolina, Chapel Hill

DAVID W. K. ACHESON, Center for Food Safety and Applied Nutrition, Food and Drug Administration, Rockville, Maryland

RUTH L. BERKELMAN, Emory University, Center for Public Health Preparedness and Research, Rollins School of Public Health, Atlanta, Georgia

ROGER G. BREEZE, Centaur Science Group, Washington, DC

STEVEN J. BRICKNER, Pfizer Global Research and Development, Pfizer Inc., Groton, Connecticut

ENRIQUETA C. BOND, Burroughs Wellcome Fund, Research Triangle Park, North Carolina

NANCY CARTER-FOSTER, Program for Emerging Infections and HIV/AIDS, U.S. Department of State, Washington, DC

GAIL H. CASSELL, Eli Lilly & Company, Indianapolis, Indiana

BILL COLSTON, Lawrence Livermore National Laboratory, Livermore, California

RALPH L. ERICKSON, Global Emerging Infections Surveillance and Response System, Department of Defense, Silver Spring, Maryland

MARK B. FEINBERG, Merck Vaccine Division, Merck & Co., West Point, Pennsylvania

J. PATRICK FITCH, National Biodefense Analysis and Countermeasures Center, Frederick, Maryland

DARRELL R. GALLOWAY, Medical S&T Division, Defense Threat Reduction Agency, Ft. Belvoir, Virginia

S. ELIZABETH GEORGE, Biological and Chemical Countermeasures Program, Department of Homeland Security, Washington, DC

JESSE L. GOODMAN, Center for Biologics Evaluation and Research, Food and Drug Administration, Rockville, Maryland

EDUARDO GOTUZZO, Instituto de Medicina Tropical–Alexander von Humbolt, Universidad Peruana Cayetano Heredia, Lima, Peru

JO HANDELSMAN, College of Agricultural and Life Sciences, University of Wisconsin, Madison

CAROLE A. HEILMAN, Division of Microbiology and Infectious Diseases, National Institute of Allergy and Infectious Diseases, National Institutes of Health, Bethesda, Maryland

DAVID L. HEYMANN, Polio Eradication, World Health Organization, Geneva, Switzerland

PHIL HOSBACH, New Products and Immunization Policy, Sanofi Pasteur, Swiftwater, Pennsylvania

JAMES M. HUGHES, Global Infectious Diseases Program, Emory University, Atlanta, Georgia
STEPHEN A. JOHNSTON, Arizona BioDesign Institute, Arizona State University, Tempe
GERALD T. KEUSCH, Boston University School of Medicine and Boston University School of Public Health, Massachusetts
RIMA F. KHABBAZ, National Center for Preparedness, Detection and Control of Infectious Diseases, Centers for Disease Control and Prevention, Atlanta, Georgia
LONNIE J. KING, Center for Zoonotic, Vectorborne, and Enteric Diseases, Centers for Disease Control and Prevention, Atlanta, Georgia
GEORGE W. KORCH, United States Army Medical Research Institute for Infectious Diseases, Ft. Detrick, Maryland
JOSHUA LEDERBERG, The Rockefeller University, New York
LYNN G. MARKS, Medicine Development Center, GlaxoSmithKline, Collegeville, Pennsylvania
EDWARD McSWEEGAN, National Institute of Allergy and Infectious Diseases, National Institutes of Health, Bethesda, Maryland
STEPHEN S. MORSE, Center for Public Health Preparedness, Columbia University, New York
MICHAEL T. OSTERHOLM, Center for Infectious Disease Research and Policy, School of Public Health, University of Minnesota, Minneapolis
GEORGE POSTE, Arizona BioDesign Institute, Arizona State University, Tempe
DAVID A. RELMAN, Stanford University, Palo Alto, California
GARY A. ROSELLE, Central Office, Veterans Health Administration, Department of Veterans Affairs, Washington, DC
JANET SHOEMAKER, Office of Public Affairs, American Society for Microbiology, Washington, DC
BRIAN J. STASKAWICZ, Department of Plant and Microbial Biology, University of California, Berkeley
TERENCE TAYLOR, International Council for the Life Sciences, Washington, DC

Staff

EILEEN CHOFFNES, Forum Director
ALLISON BRANTLEY, Sr. Program Assistant
KIM LUNDBERG, Research Associate[1]
ALISON MACK, Science Writer
KATE SKOCZDOPOLE, Research Associate

[1]Until November 2006.

BOARD ON GLOBAL HEALTH

Reviewers

This report has been reviewed in draft form by individuals chosen for their diverse perspectives and technical expertise, in accordance with procedures approved by the NRC's Report Review Committee. The purpose of this independent review is to provide candid and critical comments that will assist the institution in making its published report as sound as possible and to ensure that the report meets institutional standards for objectivity, evidence, and responsiveness to the study charge. The review comments and draft manuscript remain confidential to protect the integrity of the deliberative process. We wish to thank the following individuals for their review of this report:

John C. Bailar III, University of Chicago, Illinois, Professor Emeritus
David P. Fidler, Indiana University School of Law, Bloomington Campus
Stephen S. Morse, Center for Public Health Preparedness, National Center for Disaster Preparedness and Columbia University Mailman School of Public Health, New York
Mary E. Wilson, Department of Population and International Health, Harvard University

Although the reviewers listed above have provided many constructive comments and suggestions, they were not asked to endorse the final draft of the report before its release. The review of this report was overseen by **Melvin Worth**, Scholar-in-Residence, Institute of Medicine. Appointed by the National Research Council, he was responsible for making certain that an independent examination of this report was carried out in accordance with institutional procedures and that all review comments were carefully considered. Responsibility for the final content of this report rests entirely with the authoring committee and the institution.

Preface

The Forum on Emerging Infections was created by the Institute of Medicine (IOM) in 1996 in response to a request from the Centers for Disease Control and Prevention (CDC) and the National Institutes of Health (NIH). The purpose of the Forum is to provide structured opportunities for leaders from government, academia, and industry to meet and examine issues of shared concern regarding research, prevention, detection, and management of emerging or reemerging infectious diseases. In pursuing this task, the Forum provides a venue to foster the exchange of information and ideas, identify areas in need of greater attention, clarify policy issues by enhancing knowledge and identifying points of agreement, and inform decision makers about science and policy issues. The Forum seeks to illuminate issues rather than resolve them; for this reason, it does not provide advice or recommendations on any specific policy initiative pending before any agency or organization. Its value derives instead from the diversity of its membership and from the contributions that individual members make throughout the activities of the Forum. In September 2003, the Forum changed its name to the Forum on Microbial Threats.

ABOUT THE WORKSHOP

The failure to develop an effective plan before the occurrence of a public health emergency can have devastating consequences, as was recently demonstrated by the federal, state, and local responses to hurricanes Katrina and Rita. The National Response Plan, developed by the Department of Homeland Security in December 2004, serves as the blueprint for the coordination—such as there is—of federal agencies during any emergency. Additionally, the World

Health Organization created a pandemic preparedness plan early last year, and the Department of Health and Human Services released a more specific pandemic influenza plan in November 2005. At the same time, some countries have created influenza contingency plans, and many states in this country are now doing the same.

Local and state governments share the responsibility for protecting their citizens from disasters and for helping them recover when a disaster strikes. In some cases, a disaster is beyond the capabilities of the state and local governments to handle. The Stafford Disaster Relief and Emergency Assistance Act (1988), as amended, establishes a process for requesting and obtaining a Presidential disaster declaration, defines the type and scope of assistance available from the federal government, and sets the conditions for obtaining assistance. The Federal Emergency Management Agency (FEMA) Directorate, now part of the Department of Homeland Security, is given the task of coordinating the response.

In the fall of 2005, the federal government's response to hurricanes Katrina and Rita in support of state and local government first responders was less than optimal, despite a lead time of at least 96 hours. In the case of an infectious disease emergency—such as the pandemic spread of H5N1 avian influenza—the roles and responsibilities of local, state, and federal first responders will need to be defined in advance in order to assure an effective response that could lessen the morbidity and mortality associated with such an event.

Even though governments at all levels are beginning to formulate public health emergency response plans, the critical ethical and legal issues involved in implementing these plans and communicating these plans to the public in a transparent fashion are often pushed to the side. Past public health emergencies—including influenza pandemics, biological threats and terrorism, SARS, and the ongoing HIV/AIDS crisis—have offered numerous lessons that can be applied in future infectious disease outbreaks.

To examine the ethical and legal aspects of preparing for pandemic disease, the IOM's Forum on Microbial Threats hosted a public workshop on September 19-20, 2006, in Washington, DC. The presentations and discussions of the workshop were intended to explore the existing knowledge and unanswered questions pertaining to (but not limited to) the following topics:

- Understanding the Challenges of the Future by Examining the Past: Influenza/Smallpox/SARS
- Domestic, Regional, and International Preparedness Planning
- Disease Intervention Strategies—Quarantine, Containment, and Modeling
- Priority Setting for Access to Limited-Availability Health Care Resources

ACKNOWLEDGMENTS

The Forum on Microbial Threats and the IOM wish to express their warmest appreciation to the individuals and organizations who gave their valuable time to provide information and advice to the Forum through their participation in this workshop. A full list of presenters can be found in Appendix A.

The Forum also would like to express its deepest appreciation to Dr. Mirta Roses Periago, director of the Pan American Health Organization (PAHO), for allowing the Forum to hold this workshop at the PAHO headquarters in Washington, DC. Special thanks and gratitude are also extended to the tireless efforts of Ed Harkness, Rickey Harpster, Sergio Chacon-Oportus, Freddie Aviless, Orlando Ortiz, and, Amira Nikolas without whose help this workshop would not have been possible.

The Forum is indebted to the IOM staff who contributed during the course of the workshop and the production of this workshop summary. On behalf of the Forum, we gratefully acknowledge the efforts led by Eileen Choffnes, director of the Forum, and Kate Skoczdopole, research associate, for dedicating much effort and time to developing this workshop's agenda and for their thoughtful and insightful approach and skill in translating the workshop's proceedings and discussion into this workshop summary. We would also like to thank the following IOM staff and consultants for their valuable contributions to this activity: Patrick Kelley, Alison Mack, Bronwyn Schrecker, Allison Brantley, Kim Lundberg, Lara Andersen, Kim Weingarten, Angela Mensah, Dalia Gilbert, Thelma Cox, and Robert Pool.

Finally, the Forum wishes to recognize the sponsors that supported this activity. Financial support for this project was provided by the U.S. Department of Health and Human Services: National Institutes of Health, National Institute of Allergy and Infectious Diseases, Centers for Disease Control and Prevention, and Food and Drug Administration; U.S. Department of Defense: Global Emerging Infections Surveillance and Response System, Walter Reed Army Institute of Research, and Defense Threat Reduction Agency; U.S. Department of Veterans Affairs; U.S. Department of Homeland Security; U.S. Department of State; Lawrence Livermore National Laboratory; American Society for Microbiology; Sanofi Pasteur; Burroughs Wellcome Fund; Pfizer; GlaxoSmithKline; Infectious Diseases Society of America; and the Merck Company Foundation. The views presented in this workshop summary report are those of the workshop participants and rapporteurs and are not necessarily those of the Forum on Microbial Threats or its sponsors.

Stanley M. Lemon, *Chair*
P. Frederick Sparling, *Vice-chair*
Margaret A. Hamburg, *Vice-chair*
Forum on Microbial Threats

Contents

Tables, Figures, and Boxes

TABLES

FIGURES

BOXES

ANNEX

Summary and Assessment

ETHICAL AND LEGAL CONSIDERATIONS IN MITIGATING PANDEMIC DISEASE

In recent public workshops and working group meetings, the Forum on Microbial Threats of the Institute of Medicine (IOM) has examined a variety of infectious disease outbreaks with pandemic potential, including those caused by influenza (IOM, 2005) and severe acute respiratory syndrome (SARS) (IOM, 2004). Particular attention has been paid to the potential pandemic threat posed by the H5N1 strain of avian influenza, which is now endemic in many Southeast Asian bird populations. Since 2003, the H5N1 subtype of avian influenza has caused 185 confirmed human deaths in 11 countries, including some cases of viral transmission from human to human (WHO, 2007). But as worrisome as these developments are, at least they are caused by known pathogens. The next pandemic could well be caused by the emergence of a microbe that is still unknown, much as happened in the 1980s with the emergence of the human immunodeficiency virus (HIV) and in 2003 with the appearance of the SARS coronavirus.

Previous Forum meetings on pandemic disease have discussed the scientific and logistical challenges associated with pandemic disease recognition, identification, and response. Participants in these earlier meetings also recognized the difficulty of implementing disease control strategies effectively and, at the

The Forum's role was limited to planning the workshop, and the workshop summary has been prepared by the workshop rapporteurs as a factual summary of what occurred at the workshop.

same time, with fairness and justice. Many of the proposed disease mitigation strategies may have unintended—and often undesirable—consequences, such as adverse economic effects or the restriction of civil rights and civil liberties. To focus attention on these concerns as well as on other profound ethical and legal issues that are inherent in various pandemic disease mitigation approaches being proposed domestically and internationally, the Forum convened a public workshop, *Ethical and Legal Considerations in Mitigating Pandemic Disease* on September 19–20, 2006. Through invited presentations and discussions, participants explored lessons learned from past pandemics, identified barriers to equitable and effective responses to future pandemics, and examined opportunities to overcome these obstacles through research, policy, legislation, communication, and community engagement.[1]

ORGANIZATION OF THE WORKSHOP SUMMARY

This workshop summary was prepared for the Forum membership in the name of the rapporteurs and includes a collection of individually-authored papers and commentary. Sections of the workshop summary not specifically attributed to an individual reflect the views of the rapporteurs and not those of the Forum on Microbial Threats, its sponsors, or the IOM. The contents of the unattributed sections are based on the presentations and discussions that took place during the workshop.

The workshop summary is organized into chapters as a topic-by-topic description of the presentations and discussions that took place at the workshop. Its purpose is to present lessons from relevant experience, delineate a range of pivotal issues and their respective problems, and to offer some potential responses as described by the workshop participants.

Although this workshop summary provides an account of the individual presentations, it also reflects an important aspect of the Forum philosophy. The workshop functions as a dialogue among representatives from different sectors and presents their beliefs about which areas may merit further attention. The reader should be aware, however, that the material presented here expresses the views and opinions of the individuals participating in the workshop and not the deliberations of a formally constituted IOM study committee. These proceedings summarize only what participants stated in the workshop and are not intended to be an exhaustive exploration of the subject matter or a representation of consensus evaluation.

[1]While many of the papers within this report focus on pandemic influenza, these observations may apply to any pandemic, as recognized by David Heymann in his keynote address.

LESSONS FROM THE PAST

Current thinking on the prevention and control of pandemic disease is informed, to a large extent, by the past century's experiences with emerging, reemerging, and novel infectious disease threats. A careful examination of community responses to a broad range of infectious diseases, however, reveals enduring dilemmas that must be addressed anew with each pandemic threat. Workshop participants discussed a range of legal and ethical issues that were raised by the great influenza pandemic of 1918; the recent and mercifully short-lived outbreak of severe acute respiratory syndrome (SARS); the ongoing major epidemics such as HIV/AIDS and endemic diseases such as malaria; the efforts to control and eliminate poliovirus; the singular triumph over smallpox; and the threatened 1976 "swine flu" pandemic that wasn't.[2]

In his keynote address (see Chapter 1), David Heymann, Executive Director for Communicable Diseases at the World Health Organization (WHO), discussed notable outbreaks of emerging and reemerging diseases in the context of several interrelated issues, each of which—either individually or collectively—carries profound ethical and legal implications:

- The vulnerability of health workers to infectious disease and their duty to provide care
- Each country's responsibility to reduce the international spread of infectious diseases while simultaneously preserving trade
- Ensuring equitable access to health-care resources
- Balancing individual rights and the public good

Defined by past epidemics, these issues challenge our highly interconnected world with unprecedented urgency, as we anticipate the next infectious disease pandemic.

In response to this challenge, WHO has been engaged in a process of revising

[2]In that year, several soldiers stationed at Fort Dix, New Jersey contracted a novel respiratory virus, resulting in one death. Upon discovery that the victims were infected with a swine influenza virus, the Centers for Disease Control (CDC)—despite uncertainty as to the transmissibility of the virus—recommended the mass vaccination of the U.S. population. After the federal government agreed to indemnify vaccine manufacturers against claims of adverse reactions, 150 million doses of vaccine were produced. However, when five months passed without a single reported case of influenza, demand for vaccination dwindled (although it was briefly revived due to an outbreak of what proved to be Legionnaire's Disease). The immunization campaign was finally suspended after the vaccine was linked with Guillain-Barré syndrome (GAO, 2000, 1977; ASTHO, 2002; Sencer and Millar, 2006; Neustadt and Fineberg, 1978).

the International Health Regulations (IHR).[3] Heymann described recent efforts to expand the concept of "reportable diseases" (originally limited to cholera, plague, and yellow fever) to encompass all infectious diseases of global importance, including emerging microbial threats such as SARS (see below). The revised IHR will come into full effect in July 2007, but at the request of the World Health Assembly, WHO is prepared to implement the revisions sooner if an influenza pandemic should materialize before that date.

The Pandemic Response in History

Viewing the lessons of past pandemics from a historian's perspective, speaker Howard Markel of the University of Michigan described the organization of common social responses to epidemic or pandemic disease into narrative frameworks (see Markel, page 44). One such model, described by Charles Rosenberg, portrays an epidemic as a drama in four acts (Rosenberg, 1992):

- "Progressive revelation," in which members of a community begin to acknowledge casualties resulting from the spread of a particular contagious disease
- "Managing randomness," in which community members seek explanations (often religious ones) for the seeming arbitrariness of infection
- "Negotiating public response," in which community members demand collective action
- "Subsidence and retrospection," often leading to complacency as the memory of the epidemic fades from the community

Markel's own analysis of the social responses to epidemic or pandemic disease identifies a series of central themes, or leitmotivs, that have recurred since the Black Death (bubonic plague) of the Middle Ages:

- The public's understanding about how a disease is transmitted will affect the course of the epidemic.
- The economic consequences of an epidemic influence the public's response to the crisis.
- The extent and speed of travel of both people and goods are major factors in the spread of pandemic disease locally and globally.
- Microbes that kill relatively few people, but do so quickly and spectacu-

[3]In 1969, WHO changed the name of the International Sanitary Regulations to the International Health Regulations, and made some substantive changes at that time. The International Sanitary Regulations are direct descendents of the international sanitary conventions adopted from the 1890s through the 1940s.

larly (e.g., Ebola hemorrhagic fever, anthrax) get more attention than ongoing pandemics that kill millions year after year (e.g., tuberculosis, HIV/AIDS).

- Media coverage, which can both inform and misinform the public, influences the course of an epidemic.
- Governments will often attempt to conceal outbreaks from the world at large, typically in an effort to protect economic assets and trade.
- "Undesirable" social groups may be blamed for an epidemic or unfairly treated in the name of preventing disease transmission.

Escaping the 1918 Pandemic

In addition to offering insights into the legal and ethical issues afforded by a review of past infectious disease pandemics, Markel described how certain historical data—in this case, evaluating the effectiveness of past intervention efforts—can inform efforts today to reduce infectious disease transmission and its impact. As Markel pointed out, American communities produced a vast body of information between 1918 and1920 on the use of what are now termed "nonpharmaceutical interventions" (NPIs) against pandemic influenza. A critical analysis of these data may contain insights for mitigating the impacts of future pandemics.

By 1918, science was sufficiently sophisticated to characterize the most lethal infectious outbreak in recorded history and even to anticipate that such an event would occur (IOM, 2005). Nevertheless, between 50 and 100 million people perished in the global pandemic that began that year, many of them young adults. Today, although substantial efforts are under way to develop and stockpile vaccines and antiviral pharmaceuticals in anticipation of a threatened pandemic of H5N1 avian influenza, it is unlikely that enough of these measures will be available at the beginning of such a crisis, and their effectiveness is far from assured. Should influenza—or any other infectious disease with pandemic potential—strike, the initial individual and community containment measures employed will almost certainly be similar to the nonpharmaceutical interventions that were used almost a century ago: isolation of ill persons and quarantine of their suspected contacts from the "well"; social distancing; simple sanitary measures such as washing hands and wearing facemasks; and providing the public with information about the disease and its risks (Markel et al., 2006).

In his contribution to this volume, Markel observed that NPIs are generally considered to have offered little, if any, protection against the severe and fast-spreading influenza of 1918. While he conceded that "no systematic study exists on the relationship, positive or negative, between influenza case incidence and death rates during the 1918 pandemic and the various NPIs put into effect by the most-populated urban centers in the United States," he observed that certain combinations of NPIs may have lowered death rates in some communities in the United States. His study of "escape communities"—an ongoing examination of

NPIs used in American cities in 1918—suggests that stringent sequestration measures, applied well in advance of influenza's arrival and kept in place for extended periods, was associated with reduced influenza mortality.

While NPIs may indeed have contained influenza in some communities in 1918-1920, Markel acknowledged that implementing similar strategies in the U.S. today would present a different set of challenges. For instance, the notion that the right to civil liberty should influence public health policy—particularly with respect to minorities and the poor—is a very recent concept, he said: "The idea that you would be planning a public health policy with cultural or ethical or social or legal implications, as . . . we are discussing today . . . is quite new in the history of epidemics and medicine."

Learning from SARS

The emergence of SARS in November 2002 gave the world an opportunity to respond to a controllable human pandemic, thanks, in large part, to the limited transmissibility of the SARS coronavirus, particularly in the early stages of the illness (IOM, 2004). The four ethical issues introduced by Heymann (see above) are prominent within this story, which he recounts in detail in Chapter 1. This experience demonstrated the feasibility of containing a pandemic through a coordinated, international effort, but it also highlighted the economic and political costs associated with reporting an outbreak and their potential to undermine efforts to protect the global community from infectious diseases.

Heymann commented that the process by which the WHO detected and responded to SARS represented an important milestone. The response to SARS gave priority to global public health over national sovereignty, and it challenged national control of communications and public health activities (Heymann, 2006a; Fidler, 2004). The experience spurred efforts to establish ethical and legal guidelines for international cooperation and collaboration through the aforementioned revision of the IHR (Gostin, 2004; Fidler and Gostin, 2006).

The revised IHR has been welcomed as a first step toward a much-needed comprehensive global plan for addressing infectious disease. However, some members of the public health community—including some workshop participants—have expressed concerns that the regulations will not be completely effective because they do little to address the economic barriers to reporting infectious disease (Cash and Narashimhan, 2000; Fidler and Gostin, 2006). These critics assert that international efforts must also compensate countries for the costs of reporting and containing infectious disease outbreaks, ensure that trade and travel restrictions are cost-effective, and support public health capacity-building in vulnerable developing countries. An example of this sort of effort is the recent U.S. decision to place part of its pre-pandemic vaccine and antiviral stockpile close to vulnerable populations in Southeast Asia.

Participants also discussed the possibility of enforcing the IHR's infectious disease reporting conventions by linking them to participation in international trade (e.g., through the World Trade Organization). While everyone acknowledged the critical influence that trade has on outbreak reporting, some expressed doubts that such global governance would be acceptable to most nations; the United States, for example, has refused to sign similar conventions that it believed might compromise its sovereignty. Heymann described efforts to end the imposition of ill-founded trade embargoes as an example of how WHO, in partnership with groups such as the Food and Agriculture Organization (FAO) of the United Nations and the World Organization for Animal Health (OIE[4]), is attempting to increase acceptance and improve enforcement of the IHR in the global community.

An Exceptional Case: Smallpox Eradication

In 1980, smallpox became the first—and, to date, only—human infectious disease to be eradicated from the planet. Speaker D.A. Henderson, who led the quarter-century campaign to eliminate the disease, explained that the success of WHO's smallpox eradication program resulted in part from lessons learned from the failure of a similar campaign to eliminate malaria, which also began in the mid-1950s (Henderson, 1999). Applying these hard-learned lessons, the smallpox program took advantage of available resources in host countries, adopted broad goals that could be achieved in multiple ways, and supported a wide range of clinical, epidemiological, and operational research. Eradication of the smallpox virus was also favored by several important and unique aspects of the disease and the agent itself, including among others, the high rate of symptoms among the infected; the efficacy of the smallpox vaccine; and the absence of a nonhuman host for the virus.

Henderson also explored a series of ethical issues raised by the smallpox eradication campaign. He noted, for example, that advocates of eradication consider it to be an important element of distributive justice, since the benefits of vaccination extend to all members of a community. On the other hand, eradication also raises the possibility that individual rights will be compromised if mandatory vaccination becomes necessary. Henderson also observed that "top-down" disease eradication programs may compete for resources with "bottom-up" basic health initiatives. In the case of smallpox, he argued, providing vaccination throughout the community also served the needs of basic health services, particularly since the campaign provided a model for vaccinating against other important diseases.

Indeed, the eradication of smallpox gave birth to a new infectious disease management paradigm, and by 1990 immunization programs inspired by the success of the smallpox campaign had vaccinated 80 percent of the world's children

[4]Office International des Epizooties.

against six major diseases: tuberculosis, diphtheria, pertussis, tetanus, measles, and polio. Such new approaches represent, in Henderson's words, "key steps in revolutionizing and revitalizing public health."

PANDEMIC PREPAREDNESS PLANNING

In their workshop presentations, representatives of the U.S. Department of Health and Human Services (HHS) and the Pan American Health Organization (PAHO) described the efforts that are now being made to prepare for pandemic influenza in the United States and in Latin America and the Caribbean (see Gellin summary and Mujica et al. in Chapter 2). From subsequent discussions it was clear that there is a growing awareness among pandemic planners of the magnitude of the task at hand as well as of the scarcity of resources that are available to accomplish it. In particular, as described below, a good deal of attention has been paid to planning for and mitigating the critical shortages that are expected in countermeasures, medical care, and essential community services. Furthermore, as workshop participants discussed, the distribution of vital but limited resources raises a number of practical, ethical, and legal issues that must be considered in planning for a pandemic.

Influenza Pandemic Planning in the Americas

Oscar Mujica, communicable disease epidemiologist for the Pan American Health Organization (PAHO), described his organization's efforts to plan for pandemic influenza in Latin America and the Caribbean, in collaboration with partners that include the U.S. Agency for International Development (USAID), the Centers for Disease Control and Prevention (CDC), the World Bank, the Food and Agriculture Organization (FAO), and the United Nations Children's Fund (UNICEF). In order to pursue its primary goal of strengthening pandemic preparedness in each country within the region, PAHO has conducted a series of planning workshops for national representatives, followed by tabletop simulation exercises (see Mujica et al., page 66). Guided by two key WHO documents, the Global Influenza Preparedness Plan WHO, 2005a) and the Checklist for Influenza Pandemic Preparedness Planning (WHO, 2005b), the PAHO planning workshops have addressed a series of legal issues (such as establishing a legal basis for travel restrictions, isolation, and quarantine) and ethical issues (such as access to scarce resources, compulsory vaccination, and movement restrictions) as well as other strategic and pragmatic concerns.

Countries in the Americas have made measurable progress in advancing their preparedness plans for pandemic influenza, Mujica reported, and the recognition of the likely social and economic consequences of pandemic disease has led to increased intersectoral collaboration. Mujica also said he had observed a tendency toward greater acceptance of the IHR, as national governments gained a greater appreciation of how global disease transmission and trade are interconnected. However, as Mujica noted, considerable challenges must still be overcome before the region fully implements its pandemic plans. This is particularly true at the local level, where health-care resources are often severely constrained.

The U.S. Strategy for Pandemic Influenza

Bruce Gellin, director of the National Vaccine Program Office at HHS, introduced that agency's pandemic influenza plan as one example of the many such plans under development at the federal level in the United States (HHS, 2006). Agency-specific pandemic influenza plans fit into the more general framework of the National Strategy for Pandemic Influenza (NSPI) created by the Homeland Security Council (Homeland Security Council, 2005, 2006). A detailed NSPI implementation plan, released in May 2006, is intended to guide all federal departments and agencies—as well as non-federal entities such as state and local governments and the private sector—as they define and carry out actions to address a potential pandemic (see Gellin summary, pages 61-65).

A critical element of the NSPI is its potential for adaptation to a broad range of pandemic scenarios. Based on the six-phase pandemic scale developed by WHO (WHO, 2005a), the NSPI is organized into "response stages" defined by events in a developing pandemic (e.g., a confirmed human outbreak overseas). It sets out a series of appropriate goals, actions, and policy decisions for each stage. The resulting catalog of more than 300 possible actions to be taken by federal departments and agencies also includes progress measures and timelines for putting each action into effect. The specific actions taken in an actual pandemic would be tailored to fit its epidemiological and sociopolitical characteristics, Gellin explained, but the NSPI implementation plan is intended to cover a broad range of contingencies.

Reflecting on the process of developing a national pandemic strategy, workshop participants applauded efforts to date but voiced concerns regarding the translation of national policies into local actions. Asserting that all responses to pandemic disease are local, Steven Bice, a former CDC infectious disease specialist now at Battelle Science and Technology International, advised planners to consult with the public health officials "who will have the fight on their shoulders" in a pandemic. The current plan, he said, represents "a lot of top-down planning and not a lot of listening up . . . and if I could recommend one thing above all others, [it would be to] start listening very, very carefully to the states and the locals." State and local officials, Bice said, want guidance from the federal government on pandemic planning, such as feedback on the results of tabletop exercises or drills to test state and local preparedness. He further noted that the state and local perspective is often lacking in the evaluation of such exercises at the federal level. During the discussion, one of the meeting's participants affiliated with a state public health department encouraged a higher level of engagement between federal pandemic planners and state public health officials.

Addressing Shortages: Vaccines

Gellin's own HHS program is charged with the dual tasks of creating pre-pandemic vaccine stockpiles and developing plans for vaccine access, administration, and distribution. Given the high likelihood that vaccine supplies would be limited at best in an influenza pandemic, his office is also charged with setting priorities for the distribution of limited supplies of vaccine. Gellin described the current recommendations for pandemic vaccine prioritization (see Table SA-1), with the caveat that these priorities remain under review and, in the event of a pandemic, would be subject to change based on exactly how the disease spreads in space and time.

As is the case for any distribution of scarce resources, the assignment of priorities for influenza vaccination involves, in addition to scientific and admin-

TABLE SA-1 NVAC/ACIP Recommendations for Prioritization of Pandemic Influenza Vaccine

Tier 1A	Health-care workers • Health-care workers with direct patient contact and critical health-care support staff • Vaccine and antiviral manufacturing personnel
Tier 1B	Highest-risk groups • Patients 65 and older with at least one high-risk condition • Patients 6 months to 64 years with at least two high-risk conditions • Patients hospitalized in the past year because of pneumonia, influenza, or other high-risk condition
Tier 1C	Household contacts and pregnancy • Household contacts of children under 6 months • Household contacts of severely immunocompromised individuals • Pregnant women
Tier 1D	Pandemic responders • Key government leaders and critical pandemic public health responders
Tier 2A	Other high-risk groups • Patients 65 and older with no high-risk conditions • Patients 6 months to 64 years with one high-risk condition • Children 6 months to 23 months
Tier 2B	Critical infrastructure groups • Other public health emergency responders, public safety workers, utility workers, critical transportation workers, and telecommunications workers
Tier 3	• Other key government health-care decision makers • Individuals providing mortuary services
Tier 4	• Healthy patients 2 to 64 years without any high-risk conditions

SOURCE: HHS (2005).

istrative considerations, implicit and explicit value judgments (Kotalik, 2005). Workshop participants discussed these particular judgments at some length, raising several concerns. The first involved the apparent low priority placed on support personnel who maintain the social and physical infrastructure of communities, such as police officers. Gellin readily acknowledged that this consideration was not addressed in the current vaccination priorities, which were developed from the point of view of public health. He maintained, however, that planned revisions to the vaccination priorities plan could take into account the preservation of public safety and civil order. Concerns were also raised about the dilemmas that influenza's wavelike pattern of spread over space and time would pose. While a large-scale early release of vaccine would leave little in reserve for regions experiencing a later and potentially more severe wave (as has occurred in several past influenza pandemics), withholding vaccine from people exposed to influenza during the "first wave" would be difficult to justify if people were dying.

Participants also explored the potential expansion of domestic vaccine production capacity, which could reduce the need for prioritizing access to vaccine during a pandemic (or during a shortage of seasonal vaccine, as has occurred in the United States recently). While U.S. demand for seasonal influenza vaccine—a key determinant of manufacturing capacity—has increased in recent years, barely half of all U.S. health-care workers receive flu shots. Some participants said that understanding and changing the behavior of these influential workers could be an important step toward boosting production, availability, and access to seasonal—and thus also to pandemic—influenza vaccine.

The recognition of a new generation of vaccines currently under development for use against anthrax and smallpox (and possibly influenza) led some participants to question the substantial U.S. investment in stockpiling current vaccines. Unfortunately, Gellin pointed out, it would be imprudent to wait for better vaccines before accumulating stockpiles because of the immediacy of the threat posed by the various infectious agents. Gellin also noted that as this country's experience with the 1976 "swine flu" scare demonstrated, correctly timing the switch from producing a seasonal flu vaccine to producing a strain-specific pandemic influenza vaccine is an inherently difficult task.

Addressing Shortages: Antiviral Drugs

In addition to pre-pandemic vaccines, the U.S. is also stockpiling enough of the antiviral drug oseltamivir (Tamiflu®) to treat the 25 percent of the population projected to become ill during an influenza pandemic as well as several million more doses for use in breaking the transmission cycle during a potential outbreak. Most of the antiviral stockpile is being purchased and maintained by the federal government, Gellin said; the remainder is available for states to purchase at a federally subsidized rate. Private individuals and institutions have also begun to accumulate oseltamivir, which could potentially hinder the government stockpiling efforts. As the supply of oseltamivir begins to catch up with demand, Gellin noted, discussions about its use should shift to ethical concerns regarding fair access and the potential for increased viral resistance to the drug caused either by self-medication or by the indiscriminate use of private stockpiles.

Not only might antiviral drugs be in short supply—a likely scenario if pandemic influenza strikes in the near future—but medical researchers do not know just how effective any particular antiviral drug is likely to be against pandemic influenza. For example, as Martin Cetron, director of CDC's Division of Global Migration and Quarantine, explained, oseltamivir could improve individual patient outcome if given when symptoms first appear and if the infection is caused by a strain of influenza that is sensitive to the drug, but it might still have only a modest effect on transmission. On the other hand, antiviral prophylaxis[5] may substantially reduce virus transmission by shortening the duration of viral shedding and thereby minimizing the likelihood of infection. Promoting or stockpiling oseltamivir for H5N1 avian influenza without specific evidence of the drug's efficacy against a pandemic strain would have a variety of legal and ethical implications, and Gellin stressed the importance of determining a drug's efficacy as early as possible in the course of a pandemic and then using those findings to develop a treatment protocol. Gellin also voiced concerns regarding the lack of an antiviral backup plan should oseltamivir prove ineffective against a viral strain demonstrating pandemic potential.

Addressing Shortages: Medical Care

While it is widely acknowledged that an infectious disease pandemic is likely to overwhelm the U.S. medical system, the federal government has given scant attention—and even less money—to redressing this situation. "There is a great gaping gap here," said speaker D.A. Henderson, who criticized government planners for focusing on what he believed to be "fringe things," such as stockpiling and delivering countermeasures of questionable efficacy, rather than concentrating its efforts on "a problem which we know we are going to have." He attributed the lack of progress toward addressing this critical and predictable need to poor communication between public health officials and hospital administrators, as well as between HHS and CDC.

Although individual hospitals are attempting to prepare themselves for pandemic influenza by conducting surge capacity trials, Henderson observed that few facilities are prepared to handle a worst-case scenario in which patients could exceed capacity by 30 to 40 percent. He predicted that under those conditions hospitals would begin to turn away patients, including some who desperately need care. In order to accommodate them, Henderson recommended the creation

[5]"Prophylaxis" is defined as a treatment, such as vaccination, used to prevent disease or stop its spread.

of alternate regional sites staffed by volunteer caregivers. He also noted that plans for medical care during a pandemic need to address such issues as liability, the credentialing of volunteers, nonpaying patients or patients without adequate health insurance, the cancellation of elective surgical procedures, and pandemic-associated losses in hospital revenue.

Workshop participants considered a variety of gaps that exist in pandemic preparations at the hospital level. According to one estimate, if an influenza pandemic occurred today, demand for ventilators would exceed supply by nearly 200 percent (Bartlett, 2006). Furthermore, as noted below, a transportation slow-down—which would be likely during a pandemic—would probably cause oxygen supplies to dwindle. Contingency plans and standards of care will be required in order for nurses to prescribe medications, as many recommend they should; others noted that such contingencies will need to be negotiated with third-party payers.

Several audience participants regarded human resources as the most critical of the many needs that will be unmet during a public health emergency, since such a shortage already exists under "normal" conditions. It was noted, for example, that hospitals in the District of Columbia currently needed approximately 1,000 nurses. In response, one audience participant observed that "if we are going to solve the emergency problem, we need to look at the underlying daily situation" and then urge policy makers to support the hiring of hospital staff in order to fill today's gaps as well as to anticipate tomorrow's crises.

Addressing Shortages: Global Supply Chains

Another far-reaching concern regarding the U.S. pandemic influenza strategy is its failure to recognize America's dependence on and interdependence with fast-moving global markets. Forum member Michael Osterholm observed, for example, that the vast majority of medicines in the U.S. are manufactured abroad or made from precursor materials that are manufactured abroad. Furthermore, critical supplies such as oxygen are delivered just in time to hospitals and other end-users and are therefore dependent upon fuel, which is also largely foreign in origin. The U.S. pandemic plan, in Osterholm's view, needs to reflect the importance of international trade and travel to our medical system and to the national economy generally. Gellin responded that this concern about the impact of a pandemic on a global economy is being addressed to some extent through the U.S. government's efforts toward pandemic planning with and by the private sector.

FROM PANDEMIC PLANNING TO PUBLIC HEALTH PREPAREDNESS

While workshop discussions focused on the imminent threat of pandemic influenza, many of the issues raised apply equally to preparations for public health emergencies of all kinds, from emerging infectious diseases to bioterrorism and natural disasters. Rather than address individual threats, several participants advocated the building of public health capacity to anticipate a range of potential crises.

As one Forum member observed, shifting to this broad concept of public health preparedness would require new investments and, therefore, new considerations of cost, benefit, and risk. He added that this shift in strategy should not be driven exclusively by the local health-care needs and economic priorities that have shaped our current public health system, but should instead anticipate the global consequences of local problems.

Strategies for Disease Containment

Workshop discussion on the topic of pandemic preparedness focused on medical interventions but, as many participants noted, given the present supplies of vaccines, antiviral drugs, and ventilators, nonpharmaceutical interventions are likely to dominate the public health response to an H5N1 avian influenza pandemic. Behavioral strategies for disease containment and mitigation are thus critical to pandemic planning, but they are also fraught with legal and ethical concerns. As described below, time-tested containment measures such as quarantine are being brought into the twenty-first century by computer-modeling simulations designed to find the optimum intervention in different scenarios. Behavioral strategies for disease containment must also be compatible with modern legal thinking, with its emphasis on objective standards and fair procedures.

Isolation and Quarantine in the Twenty-First Century

After centuries of use, *isolation*[6] and *quarantine*[7] persist as primary public health intervention tools in the twenty-first century because of their ability to limit the spread of disease. But today the mitigation of disease spread must be done in accordance with current beliefs about individual rights and civil liberties, Cetron observed (see Cetron and Landwirth, page 99). Cetron described how quarantine and social distancing (individual and community measures that reduce the frequency of human contact) have proved effective in reducing the transmission of disease. Numerous historical examples suggest that people tend individually and collectively to respond to infectious disease threats through social distancing, Cetron observed. In order to maximize the benefits of this behavior, such measures must be planned and instituted at the beginning of a pandemic.

Modern quarantine is carried out according to ethical principles used to guide public health interventions. These ethical principles include the need to demon-

[6]The practice of separating ill persons with contagious diseases from society.

[7]The restriction of movement by persons who are not ill but are presumed exposed to infectious disease.

strate necessity (justifiable harms); the importance of using the least restrictive means of achieving a public good; the existence of mechanisms for notification and appeal (due process); and fairness in carrying out the intervention (Upshur, 2002). As Cetron explained, quarantine can only be justified in the case of a highly dangerous, contagious disease and only for as long as is necessary to protect the public. The quarantine need not be absolute, since even a "leaky" quarantine may be effective in disease mitigation. Cetron added that pandemic-preparedness plans should guarantee that quarantined individuals will have access to needed goods and services, and that quarantine-associated stigma should be anticipated and actively discouraged.

Modeling the Effects of Social Distancing

There is limited empirical evidence supporting the effectiveness of social distancing in fighting the spread of infectious disease. As illustrated in Figure SA-1, there are several ways that social distancing—in combination with additional infection control measures—could be expected to alter the course of an outbreak: it could can delay the peak of the outbreak (#1); it could spread the outbreak out in time, thereby easing the peak burden on hospitals and the public health infrastructure (#2); and it could reduce the overall number and severity of cases (#3).

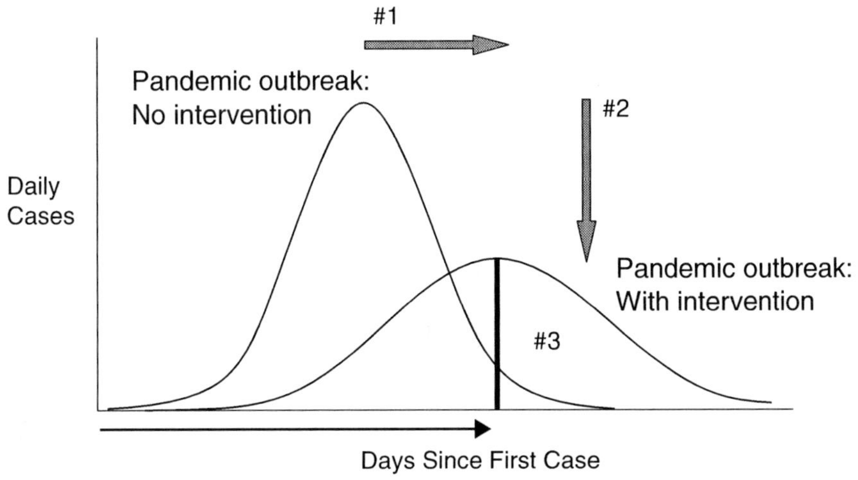

FIGURE SA-1 Community-based interventions. Figure legend: (#1) delay outbreak peak; (#2) decompress peak burden on hospitals/infrastructure; (#3) diminish overall cases and health impacts.
SOURCE: Cetron (2006).

Based on computer simulation models, specific actions that might reduce disease transmission rates include school closures; keeping children and teens at home; voluntary home isolation and quarantine; and using antiviral drugs to treat the ill and providing prophylaxis to their household contacts. As Cetron explained, these measures form part of a much broader, layered approach to behavioral intervention, which extends from individual actions (hand hygiene, cough etiquette) to global efforts (containment at the source, advisories and screening for travelers).

While social distancing measures may help slow the spread of disease, they also pose a number of other potentially far-reaching consequences, and Cetron stressed the importance of anticipating these consequences and adapting the measures accordingly. The issue of school closure is particularly contentious in this regard. While modeling indicates it to be a potent means to reduce disease transmission, its adverse consequences could be so severe and inequitable as to outweigh any benefit. D.A. Henderson of the University of Pittsburgh Medical Center cautioned against relying on models that do not take into consideration the adverse effects or practical constraints that such public health interventions would entail. Accepting such models uncritically, he warned, could result in policies that "take a perfectly manageable epidemic and turn it into a national disaster."

While agreeing that models cannot take the place of human judgment, Joshua Epstein of the Brookings Institution described how they can inform human judgment by providing insights about how different human actions might affect the course of an epidemic. In particular, he described how the reaction of people to various disease-control measures—the so-called secondary and tertiary effects of a measure—can be incorporated into a model and evaluated prior to an event. With every biological epidemic, he predicted, "there will be a secondary epidemic, a behavioral epidemic that will affect subsequent social network patterns [and] contact behavior." Epstein and coworkers have produced such a model for smallpox following a bioterrorist attack (see Epstein, page 105) and are currently developing such a model for pandemic influenza. Epstein's team is also working on a global model to demonstrate the effects of international travel restrictions and other nonpharmaceutical interventions in a variety of circumstances. And the CDC, according to Cetron, is testing a model designed to identify those behavioral measures that will save the maximum number of lives in a pandemic with the minimal economic impact. Preliminary results indicate, for example, that few additional lives will be saved by reducing the disease attack rate[8] below one percent or by shifting from voluntary to compulsory quarantine.

---

[8]The attack rate, or case rate, is a cumulative incidence rate often used for particular groups that are observed for limited periods and under special circumstances, as in an epidemic. The secondary attack rate expresses the number of cases among contacts occurring within the accepted incubation period following exposure to a primary case, in relation to the total of exposed contacts. The *infection rate* expresses the incidence of manifest and non-apparent infections (Last, 1983).

Participants noted that sound science is essential to the creation and continuous improvement of disease models. "You really have to work closely with experts in building models," Epstein advised, illustrating this point by describing the design of his group's model of smallpox transmission, in consultation with Henderson (Cummings et al., 2004). "We had intensive, regular meetings to arrive at reasoned assumptions about all the biomedical and critical behavioral aspects of this problem," Epstein recalled, adding that this process forces scientists to make their assumptions explicit and to support them, and thereby identifies gaps in knowledge. James LeDuc of CDC observed that such gaps present important opportunities for scientific research to inform ethical decision making and thereby create a foundation for public health law.

Legal Preparations for Containment

Lawrence Gostin, Director of the Center on Law and the Public's Health at Georgetown University, introduced his presentation by reviewing the legal issues surrounding nonmedical approaches to disease containment. These include the potential for infectious disease surveillance to violate individuals' rights to privacy, the economic costs of controlling zoonoses, trade limitations imposed by international travel restrictions, and the loss of personal freedom associated with community hygiene measures, such as compulsory temperature monitoring, and the spectrum of contact-reducing interventions, ranging from voluntary sheltering in place ("snow days") to mandatory isolation of the ill (see Gostin and Berkman, page 78). Given the lack of clear evidence that many of these disease containment strategies are effective, Gostin said that governments and institutions should carefully consider the possible adverse effects and unintended consequences that such measures can cause when they are designing and implementing them.

Gostin stressed the importance of legal preparedness for a pandemic, both nationally and internationally. Calling the revised IHR a "brave, bold, and innovative" move toward developing an international legal authority capable of addressing pandemic disease in a globalized society, he warned that the regulations would remain "vacant" until developing countries receive sufficient resources for compliance or until transnational health law is sufficiently strong to enforce the IHR. "We need to move toward more transnational governance," Gostin asserted, because no matter how well an individual country may prepare for a pandemic, "we are all at risk unless we have some kind of coordinated approach."

Focusing on quarantine, Gostin emphasized the importance of establishing sufficient legal authority, based on objective standards, to restrict individual freedom. If members of the public are to support such a measure, he said, they must be assured that those who are quarantined will be well taken care of—ensured a safe place to stay, provided with food and water, given adequate medical care, and so on—for the duration of the quarantine. Moreover, he contended, people subject to compulsory quarantine must have the right to a hearing to contest their confinement. It is the states rather than the federal government that have the primary legal authority to mandate quarantine measures, but the federal government has drafted the Model State Emergency Health Powers Act and Model State Health Act[9] to provide guidance to the states on how best to prepare for pandemic disease and other health crises. "Public health strategies require public trust and acceptance in accordance with the principles of social justice," Gostin concluded. "We need to remember that pandemics can be deeply divisive, and the political response profoundly reflects on the kind of society we want and aspire to be."

The Problem of Authority

Among the legal and ethical barriers to public health intervention during a pandemic, the first and most critical is the establishment of authority for decision making, according to speaker Victoria Sutton, director of the Center for Bioterrorism, Law, and Public Policy at Texas Tech University. As cruelly demonstrated in the aftermath of Hurricane Katrina, disaster response can be slowed and sometimes even brought to a halt by confusion among national, state, and local jurisdictions about who has authority and responsibility for various decisions and actions. In addition, IOM president Harvey Fineberg expressed concern that a "horizontal dimension of ambiguity" of federal public health authority has been created as a consequence of the organization of the Department of Homeland Security (DHS).

The longstanding conflict between federal and state governments concerning authority over public health is rooted, Sutton said, in decisions made early in our nation's history,[10] when disease transmission tended to be limited to communities. But because pandemic diseases are becoming recognized as posing a national threat, Sutton suggested that the federal government could pass legislation federalizing the rules for pandemic response, much as the recognition of the far-reaching and adverse effects of pollution led to federal environmental legislation. The pandemic legislation might, for example, put into effect a "cooperative federalism," in which the federal government establishes standards for implementing pandemic measures (medical care, the distribution of therapeutic countermeasures, quarantine and isolation), which are administered by the state governments and, ultimately, implemented at the local level (Sutton, 2001).

[9]See http://www.publichealthlaw.net/Resources/Modellaws.htm#MSPHPA [accessed May 17, 2007].

[10]In U.S. history, the states' rights doctrine is based on the Tenth Amendment to the Constitution, which states, "The powers not delegated to the United States by the Constitution, nor prohibited by it to the States, are reserved to the States respectively, or to the people." The term embraces both the doctrine of absolute state sovereignty that was espoused by John C. Calhoun and that of the so-called strict constructionist interpretation of the U.S. Constitution, which reserves to the state governments all powers not specifically granted by that document to the federal government. See http://www.answers.com/topic/states-rights [accessed December 27, 2006].

This model of federal leadership was endorsed by Shelley Hearne of the Johns Hopkins University, who presented a detailed strategy for building a capacity for emergency response that would protect all U.S. citizens equally (see Hearne, page 183). At present, she reported, pandemic preparedness varies greatly from state to state and from hospital to hospital, and most health-care providers and workers are not engaged in preparedness planning or implementation activities (Trust for America's Health, 2005). Hearne argued that, just as the federal government insists that each state meet certain minimal environmental standards, it should also define and uphold public health standards. Drawing a parallel with environmental legislation, she suggested that federal public health legislation might even allow citizens to file suits against the CDC or other federal agencies if public health standards were not enforced. LeDuc observed that until the distribution of public health authority is clarified and reformed, each community will have to identify and repair gaps in its own legal preparedness for a pandemic. He added that this process could be undertaken as a collaboration among federal, state, and local legal authorities. There has been some progress toward community engagement, Cetron said, thanks to efforts undertaken by the Association of State and Territorial Health Officers, the Department of Education, and the CDC, which issued guidelines on pandemic mitigation for communities in February 2007 (CDC, 2007). Feedback from the community level will continue an iterative process, intended to encourage communities to identify and independently address the challenges they will face in a pandemic. A number of state and local public health officials who attended the workshop commented that such federal guidance on pandemic mitigation is very much needed.

ETHICAL ISSUES IN PANDEMIC PLANNING AND RESPONSE

The various efforts at planning for an H5N1 avian influenza pandemic have created an awareness of the challenges involved in addressing infectious diseases in a highly interconnected and interdependent world. Participants in the many conferences, meetings, and workshops convened in response to this imminent threat have considered a broad range of potential effects and contingencies, of actions that could be taken, and of decisions that will need to be made prior to and during a pandemic. Accordingly, much of this workshop was devoted to identifying the logistical, legal, and ethical challenges that are likely to arise in such a public health emergency. Certain presentations and discussions, summarized below and in Chapters 3 and 4, focused on the creation of ethical guidelines for action in a pandemic, the engagement of the public in this process, and the maintenance of free and open communication about the process.

As Victoria Sutton noted, a fundamental question underlies the various ethical issues surrounding pandemic planning and response: Do ethics change during a pandemic? The answer, she said, is yes, just as is the case in any public emergency that involves a great deal of uncertainty and that threatens the well-being of the nation (Sutton, 2005). Sutton, who sees ethics as a precursor to law, suggested that a good definition of "pandemic flu ethic" would be "a limitation on the freedom of action or the imposition of a duty to act in the pursuit of the continued existence of life and order." This part of the workshop offered nuanced interpretations of her seemingly straightforward definition.

The Ethicist's Role

Speaking from his experience as WHO's former Director of Ethics, Trade, Human Rights and Health Law, medical ethicist Alexander Capron of the University of Southern California described how pandemic planning and response fit on a spectrum of ethical approaches (see Capron, page 157). A variety of ethical principles apply to the content of pandemic policies and the process by which such policies are established, he observed; further principles concern the individual's duty in a pandemic or focus on desired outcomes (see Table SA-2). Guided by these principles, ethicists can assist pandemic planning by recognizing and raising awareness of the values that are embedded in technical decisions and by promoting an ethical process of decision making. "The key to an ethically responsible and appropriate response is advance planning, including communication," Capron said. "Part of communication is recognizing scarcity and the resulting need for collective allocation and personal responsibility."

Capron then revisited the four primary ethical considerations raised by pandemic disease that had been discussed previously in connection with SARS and other infectious diseases (see previous section, "Learning from Pandemics Past," and Heymann, page 33):

- Access to health care
- Human rights
- Obligations of and to health-care workers
- Obligations of countries and intergovernmental organizations

Applying these ethical considerations to the current scenarios for pandemic influenza, Capron offered a number of specific issues that need to be addressed

TABLE SA-2 Variety of Ethical Approaches/Foci

- Focus of ethics can be on the content of policies (e.g., about pandemic preparations and response
- Alternatively, can focus on the process by which policies are established
- One can also incorporate both duty-oriented and outcome-oriented considerations (e.g., utility and justice) into the principles that will be used
- Ethics may coincide with prudential considerations

SOURCE: Capron (2006).

TABLE SA-3 Substantive Principles

- Principle of utility: act so as to produce the greatest good
- Principle of efficiency: minimize the resources needed to produce an objective or maximize the total benefit from a given level of resources
- Principle of fairness: treat like cases alike and avoid unfair discrimination (that is, discrimination based on irrelevant or illegitimate characteristics of a person or group)
- Principle of liberty: impose the least burden on personal self-determination necessary to achieve legitimate goals (or, broadly speaking, do not trade all freedom for security)

SOURCE: Capron (2006).

in planning for a pandemic, including the need to define standards of care that result in the fair distribution of scarce resources.

Capron also described how WHO has incorporated certain ethical principles, as outlined in Table SA-3, into technical recommendations for pandemic influenza planning and response. The CDC has taken a similar route in formulating ethical guidelines for pandemic preparation and response, according to James LeDuc (see LeDuc et al., page 90). Ethicists at that agency have suggested a number of principles, including transparency, sound science, global involvement, and procedural justice, as a foundation for legal decision making and have also identified certain ethical issues, such as allocation of countermeasures and restrictions on freedom, that are inherently part of pandemic planning and response (CDC, 2007).

Pandemics tend to produce conflicts between individual entities, either persons or nations, and groups, such as communities, countries, or intergovernmental entities, which must be resolved through "choices among goods," Capron observed. And that resolution can vary depending on the outcome of public debate and expert deliberation. Indeed, he said, rather than expecting to find an absolute "right" answer to a problem, the more viable approach is to find a solution that is both understood and accepted by the public, who ultimately must implement and abide by the decision. The public's trust must be attained by ensuring that the decision-making process, as well as the evidence used in it, is transparent and open to public debate and that the resulting decisions must be clearly articulated and justified.

Questions of Justice

In his remarks to workshop participants, Harvey Fineberg raised an additional ethical challenge: the potential conflict between preserving society as a whole and protecting its weakest members. These considerations were further explored in presentations by ethicists Ruth Faden of the Johns Hopkins University and Bernard Lo of the University of California, San Francisco. Faden's remarks focused on the well-founded expectation that the burdens of pandemic influenza—social and economic disruption, as well as morbidity and mortality—will almost certainly fall disproportionately upon the world's poorest people and countries and also upon the poorest inhabitants of wealthy countries. "The greatest moral challenge we face is how to respect commitments to social justice in the face of the overwhelming, systematic inequalities that form the backdrop for the harsh realities of pandemic flu," she said.

Faden observed that some inequalities likely to be created in an influenza pandemic are more profoundly wrong than others (see Faden, page 177). And of those most egregious inequalities, some—such as limited access to technology and barriers to communication and community engagement—represent particularly feasible targets for policy, intervention, or development. Governments, she argued, should identify opportunities to mitigate, if not eliminate, the burdens imposed by such inequalities under pandemic circumstances—and perhaps, by extension, in the broader context of public health.

There are also important practical reasons for considering the interests of the disadvantaged in the course of pandemic planning. For instance, trust in public health policies—particularly among the typically poor inhabitants of regions where an influenza pandemic is expected to begin—will invite compliance, without which infectious disease will not be able to be controlled (Bellagio Group, 2006). Lo noted in his formal remarks that in public health emergencies citizens are more willing to sacrifice self-interest in favor of the common good if they believe that everyone else is doing so; conversely, he said, people who perceive that others are receiving preferential treatment (even if this is not actually the case) are less likely to act selflessly.

Duty to Care

Health-care workers on the front lines in infectious disease outbreaks (e.g., smallpox, Ebola, and SARS) have consistently fulfilled their duty to care for patients even when it has cost them their lives (see Heymann, page 33). Ruderman and colleagues report, however, that during the SARS crisis in Canada, "serious concerns arose . . . about the extent to which health-care providers would tolerate risk of infection," leading to the anticipation of a potential crisis during a pandemic (Ruderman et al., 2006). Likewise, Sokol (2006) described defining the duty to care as an urgent and difficult task, "strewn with philosophical and logistical difficulties," that must be accomplished in order to prevent "large numbers of doctors from abandoning their patients in a crisis" (Sokol, 2006).

Capron asked whether a worker's duty to care is a product of medical training and licensure or a product of the patient's—and the community's—need for their skills (see also Chapter 4). If it is the former, Clark argued that the duty to treat overrides physician autonomy in social emergencies, even in cases that involve personal risk (Clark, 2005). If one accepts the latter premise, Capron said, anyone whose job is essential to the health-care system has an obligation to work during a pandemic, and in return, society is obliged to provide these workers with the most effective protection available. "This is part of the social contract with them, that we give them special status when it comes to prophylaxis and treatment in recognition of their role," he said.

In enforcing these dual obligations—the duty of health-care professionals to society and the reciprocal duty of society to health-care professionals—governments and professional organizations should incorporate some flexibility in the definitions of "duty to care" and "special protections," Capron said, so that they can accommodate factors such as disease-transmission dynamics and the availability of countermeasures that will be specific to each public health emergency. Furthermore, while it is possible that during a pandemic a sufficient number of health-care workers will volunteer to put themselves at risk for the benefit of the community, this cannot be guaranteed, and thus it may be necessary for states or professional organizations to mandate worker conduct. According to Capron, if such expectations for health-care workers are going to be mandated—possibly along with specific consequences for dereliction of duty—the expectations must be agreed upon as part of pandemic planning and should be done through a transparent process that involves local and national professional associations.

Ethical Guidelines for Clinicians

An influenza pandemic is likely to produce extraordinary shortages in medical care. Hospital resources—both human and material—may be stretched beyond their limits. In order to manage the many ethical dilemmas inherent in this situation, physicians and hospital administrators will need specific guidelines, Lo said (see Lo and White, page 192). His observations were echoed by several workshop participants, some of whom spoke from a personal perspective, as they themselves will be called to play certain roles in a pandemic. Among the challenges that pandemic influenza will present to clinicians, one of the likeliest and most daunting will be a grave shortage of mechanical ventilators. Such a shortage, Lo observed, will require physicians to choose which patients will receive the life-saving use of a ventilator and which will die without respiratory therapy. There will be no time to weigh alternatives in a pandemic, Lo argued, so it will be important to develop clear criteria ahead of time for when to triage patients, along with guidelines and procedures for addressing problems that will arise as the triage system is implemented, such as handling disagreements with family members and managing patients in respiratory failure who do not receive mechanical ventilation.

Lo urged pandemic planners to anticipate the ethical and legal dilemmas that doctors and other health-care providers will face in a "worst-case" ventilator shortage and to create, with input from the public and specialists in various disciplines, guidelines and procedures for dealing with shortages of ventilators and other medical supplies. While suggesting that rules for triage should maximize the number of lives saved, Lo also pointed out that physicians must rely on limited evidence to predict a patient's prognosis. Triage rules, he said, should be administered by an external authority, not the physicians dealing with the patients, and they should be implemented by physicians and other health-care workers in such a way that their fairness cannot be doubted.

Fairness in allocating scare resources will be necessary to secure public trust in the process, Lo observed, but it will not be sufficient. Triage policies and priorities must also reflect popular will, he said. Moreover, the policies must be communicated clearly and in a way that people will understand. And they must be presented in a way that leads society to accept the idea that, during an infectious disease emergency, some patients will die who might otherwise have been saved under normal circumstances. Lo also stressed the importance of providing the public with ready access to the data, reasoning, and deliberative processes that support such triage guidelines. Unfortunately, Hearne observed, some states have not only failed to engage the public in pandemic planning, but they have actively excluded them from the process and have kept their plans secret, even from hospital workers and other health-care providers.

Civic Engagement

Throughout the workshop, the public perspective dominated discussions of pandemic ethics, as reflected in presentations that advocated transparent planning processes and clear communication. But transparency is not enough, Capron said. The process should also adhere to the principle of participation, which holds that stakeholders (and who is not a stakeholder in a pandemic?) should contribute to the process of formulating objectives and adopting policies (see also Chapter 4).

As previously described, public participation has been incorporated in pandemic planning efforts undertaken by both PAHO and CDC (see Mujica et al., page 66, and Cetron and Landwirth, page 99). Faden noted that among prior attempts to engage the public in health policy making (concerning such issues as resource rationing, end-of-life care, and genetic screening), some of the most successful efforts were those that occurred on a scale that was sufficiently local that people were able to voice their individual concerns. When society wrestles with issues that affect people's daily lives, Capron said, people generally want the chance to participate in the "work of worrying." They don't always act on it, he said, "but they often miss it when they don't have it." Susan Chu, who moderates the public Internet forum FluWiki, observed that its participants exhibit a high level of engagement and sophistication. Asserting that at least "a portion of the public is teachable," Chu encouraged policy makers to speak truthfully about the possible effects of a pandemic and to enlist citizen engagement in the planning efforts.

Acknowledging that the Internet has raised the bar for participatory communications, Capron noted that public input has influenced health-care policy in many instances, perhaps most profoundly in addressing HIV/AIDS. These experiences demonstrate that public engagement produces policies that can be accepted even by those who are not satisfied by them, he said. Conversely, exclusion often creates a heated atmosphere that leads people to attack and arbitrarily reverse decisions that may be technically sound. "Things that make people angry fall apart," Capron observed.

Responding to Capron's assertion that pandemic plans should be subject to revisions based on public review of their performance, Martin Cetron predicted that engaging in this process will build community resilience, as both policy makers and the public become "acculturated to being wrong" and become familiar with uncertainty. The hoped-for result of this process, he continued, is the creation of multiple contingency plans (including criteria for implementing and suspending them) as well as mechanisms to "ensure that we are going to learn from the mistakes."

In addition to incorporating public input into pandemic planning, the federal government has been encouraging personal responsibility in emergency preparedness, and Steven Bice applauded these efforts. He briefly described the MedKit program, currently being tested by HHS, which provides families with a home supply of antimicrobial drugs as well as countermeasures for exposures to radioactive materials, which could be used in the event of a terrorist attack (CDC, 2005). Bice suggested that antiviral drugs could be provided to individuals in advance of a pandemic in much the same way, thereby providing access to medications without compromising social distancing measures.

Since the nation's experience with the aftermath of Hurricane Katrina, many Americans have come to be extremely cynical about government efforts meant to protect them from disaster, Hearne observed. As a result, she said, broad changes in public health law will be needed to prevent a potentially disastrous breakdown in public health authority during a pandemic. She expressed the hope that such reforms could be structured to engage that public in emergency planning and response, and, as a result, transform a future pandemic into a "controlled chaos event that gives us a better chance of saving more lives and having a public that trusts its government."

Ethics in the Midst of Uncertainty

While recognizing the ideal of public participation in pandemic planning, workshop participants nonetheless agreed that public health professionals must expect most people to be entirely unprepared when the next pandemic strikes. In order to mount an effective response, public health authorities will need to act rapidly and authoritatively on the basis of incomplete knowledge. This situation, which Gostin (2004) refers to as "the public health paradox," raises a host of ethical and legal issues for those who will lead the response to a threatened pandemic—scientists, public health authorities, and lawmakers (Gostin, 2004).

Fineberg warned scientists against overconfidence in assumptions based on incomplete knowledge of a few varied pandemics. "While you can be very sure that there is going to be a next time," he said, "you have to be very, very cautious" before declaring that next time to be now. In light of this uncertainty, he posed a series of questions concerning the ethical duties of biomedical experts toward political officials. The questions were the following:

- Are experts bound to frame evidence, based on their knowledge, so that politicians reach "correct" conclusions regarding a threatened pandemic?
- Should experts refrain from making conclusions and merely answer questions?
- Should experts speak directly to the media about their concerns?

Rather than offering answers, Fineberg described how various experts approached these dilemmas in the course of reacting to the appearance of "swine flu" in 1976, and he spoke of how those reactions—and their treatment in the media—shaped the nation's response to a threatened pandemic (see Chapter 4).

Finally, Fineberg said, no matter what choices are made to address a threatened pandemic—or indeed, in response to any decision of such importance—there is going to be skepticism, criticism, and differences of opinion. "There is no way to avoid the dilemmas posed by acting without full scientific knowledge," Gostin has observed (2004). In the face of such challenges, he concluded, "the only safeguard is the adoption of ethical values in formulating and implementing public health decisions."

SUMMARY OF NEEDS AND OPPORTUNITIES

This section summarizes the needs and opportunities for research and for policy making that the workshop participants mentioned frequently as being important to the development of just and ethical measures for mitigating pandemic disease.

1. A Comprehensive, Global Plan for Addressing Infectious Disease

Given the globalization of commerce, travel, and economics, and the worldwide migration of people, goods, and ideas, there needs to be a coordinated, international approach to pandemic disease mitigation which includes the following features:

- Universal implementation of core disease surveillance and control capacities
- International cooperation and coordination in disease surveillance

- Protection of international trade from unnecessary embargoes related to disease reporting, and maintaining vital global supply chains
- Compensation of citizens and governments of low-resource countries at risk for emerging infectious diseases for those sacrifices (e.g., economic consequences of disease reporting, culling of infected animals, quarantine and isolation) that benefit the global community; this would include the provision and just distribution of countermeasures, as well as economic support

2. A Pandemic Planning Process That Tackles Ethical and Legal Issues

As Alexander Capron remarked, and many others echoed, "the key to an ethically responsible and appropriate response [to pandemic disease] is advance planning." Important features of such a planning process would include, but not be limited to, the following:

- Public engagement in decision-making on issues that affect personal freedom (e.g., quarantine, school closings, and other NPIs) and the allocation of limited medical resources (see also Item #4 below)
- Adherence to ethical principles, including justifiable harms, the least restrictive means of achieving public good, procedural justice, and due process
- Interventions of proven efficacy, implemented under conditions in which their benefits to society outweigh their well-understood and potentially far-reaching consequences
- Establishment of clear authority for public health decision making during a pandemic (see Item #3 below)
- Public communication prior to—as well as during—a pandemic that explains the rationale for disease control measures, establishes realistic expectations, and allows for the "emotional rehearsal" of pandemic scenarios
- Measurable endpoints for NPIs and other disease control measures
- Research protocols to gather evidence for the efficacy of NPIs
- Capacity building to address a range of potential public health emergencies

3. Improved Coordination of Federal, State, and Local Pandemic Planning Activities

Workshop participants noted several critical weaknesses in current structures and strategies for pandemic planning. It is worth noting that the application of ethical values in formulating plans may be as, or more, important than the ethical values in the plans themselves, although both are critical elements in the planning process. Confusion as to who would be in charge of the nation's response to an unfolding pandemic is a particularly pressing concern, and it encompasses the following administrative challenges:

- Ambiguity of federal public health authority between DHS and HHS
- Longstanding conflicts between federal and state claims to public health authority, as demonstrated following hurricanes Katrina and Rita
- Insufficient engagement of state public health officials by federal pandemic planners
- Poor communication between state and local public health officials and hospital administrators

4. Public Engagement and Communication

Transparency in policy making and clear risk communication are necessary, but not sufficient, to ensure that fairness and justice prevail in the face of a threatened pandemic. Many workshop participants suggested that there was a need for a transparent, ethical decision-making process that incorporates public debate and deliberation and that has as its goal the selection of interventions that are both understood by, and acceptable to, most people. In addition, due to the inherent uncertainty of pandemic disease, preparedness plans should be "living documents" that are subjected to constant review, testing, and revision, based on evidence and experience.

REFERENCES

ASTHO (Association of State and Territorial Health Officials). 2002. *Nature's Terrorist Attack: Pandemic Influenza: Preparedness Planning for State Health Officials.* Washington, DC: Association of State and Territorial Health Officials.

Bartlett JG. 2006. Planning for avian influenza. *Annals of Internal Medicine* 145(2):141-144.

Bellagio Group. 2006 (August 25). *Bellagio Statement.* [Online]. Available: http://www.hopkinsmedicine.org/bioethics/bellagio/statement.html [accessed January 3, 2007].

Capron A. 2006 (September 19). *Session II: Domestic, Regional, and International Preparedness Planning.* Presentation at the Institute of Medicine's Forum on Microbial Threats workshop entitled "Ethical and Legal Considerations in Mitigating Pandemic Disease," Washington, DC.

Cash RA, Narashimhan V. 2000. Impediments to global surveillance of infectious diseases: Consequences of open reporting in a global economy. *Bulletin of the World Health Organization* 78(11):1358-1367.

CDC (Centers for Disease Control and Prevention). 2005. *Advisory Committee to the Director, Meeting Summary, August 25, 2005.* [Online]. Available: http://www.cdc.gov/maso/FACM/pdfs/ACDCDC/20050825%20ACDCDC%20minutes.pdf [accessed February 23, 2007].

CDC. 2007. *Ethical Guidelines in Pandemic Influenza.* Advisory Committee to the Director, Ethics Subcommittee. [Online]. Available: http://www.cdc.gov/od/science/phec/panFlu_Ethic_Guidelines.pdf [accessed April 13, 2007].

Cetron M. 2006 (September 20). *Session III: Disease Mitigation Strategies—Quarantine, Containment and Modeling.* Presentation at the Institute of Medicine's Forum on Microbial Threats workshop entitled "Ethical and Legal Considerations in Mitigating Pandemic Disease," Washington, DC.

Clark CC. 2005. In harm's way: AMA physicians and the duty to treat. *Journal of Medicine and Philosophy* 30(1):65-87.

Cummings DAT, Chakravarty S, Singa RM, Burke DS, Epstein JM. 2004. *Toward a Containment Strategy for Smallpox Bioterror: An Individual-Based Computational Approach.* Washington, DC: Brookings Institution Press.

Fidler DP. 2004. *SARS, Governance and the Globalization of Diseases.* New York: Palgrave Macmillan. Pp. 50-60.

Fidler DP, Gostin LO. 2006. The new international health regulations: An historic development for international law and public health. *Journal of Law, Medicine and Ethics* 34(1):85-94.

GAO (General Accounting Office). 1977. *The Swine Flu Program: An Unprecedented Venture in Preventive Medicine.* GAO/HRD-77-115. Washington, DC: General Accounting Office.

GAO. 2000. *Pandemic Influenza: Plan Needed for Federal and State Response.* GAO-01-4. Washington, DC: General Accounting Office.

Gostin LO. 2004. Pandemic influenza: Public health preparedness for the next global health emergency. *Journal of Law, Medicine and Ethics* 32(4):565-573.

Henderson DA. 1999. Eradication: Lessons from the past. *Morbidity and Mortality Weekly Report* 48(SU01):16-22.

Heymann DL. 2006a (June 6). The sovereignty of disease. *YaleGlobal.* [Online]. Available: http://yaleglobal.yale.edu/display.article?id=7518 [accessed January 3, 2007].

Heymann DL. 2006b (September 19). *Emerging Infectious Diseases: Past is Prologue.* Keynote address at the Institute of Medicine's Forum on Microbial Threats workshop entitled "Ethical and Legal Considerations in Mitigating Pandemic Disease," Washington, DC.

HHS (Department of Health and Human Services). 2005. Appendix D in *Pandemic Influenza Plan.* Washington, DC: Department of Health and Human Services. [Online]. Available: http://www.hhs.gov/pandemicflu/plan/appendixd.html [accessed December 27, 2006].

HHS. 2006 (June 29). *Pandemic Planning Update II.* [Online]. Available: http://www.pandemicflu.gov/plan/pdf/panflureport2.pdf [accessed January 3, 2007].

Homeland Security Council. 2005 (November). *National Strategy for Pandemic Influenza.* [Online]. Available: http://www.whitehouse.gov/homeland/nspi.pdf [accessed January 3, 2007].

Homeland Security Council. 2006 (May). *National Strategy for Pandemic Influenza Implementation Plan.* [Online]. Available: http://www.whitehouse.gov/homeland/nspi_implementation.pdf [accessed January 3, 2007].

IOM (Institute of Medicine). 2004. *Learning from SARS: Preparing for the Next Disease Outbreak.* Washington, DC: The National Academies Press.

IOM. 2005. *The Threat of Pandemic Influenza: Are We Ready?* Washington, DC: The National Academies Press.

Kotalik J. 2005. Preparing for an influenza pandemic: Ethical issues. *Bioethics* 19(4):422-431. [Online]. Available: http://www.blackwell-synergy.com/doi/pdf/10.1111/j.1467-8519.2005.00453.x [accessed January 3, 2007].

Last JM, Ed. 1983. *A Dictionary of Epidemiology.* Oxford, UK: Oxford University Press.

Markel H, Stern AM, Navarro JA, Michalsen JR. 2006. *A Historical Assessment of Nonpharmaceutical Disease Containment Strategies Employed by Selected U.S. Communities During the Second Wave of the 1918-1920 Influenza Pandemic.* Ft. Belvoir, VA: Defense Threat Reduction Agency.

Neustadt RE, Fineberg HV. 1978. *The Swine Flu Affair: Decision-Making on a Slippery Disease.* Washington, DC: Department of Health, Education, and Welfare.

Rosenberg C. 1992. What is an epidemic? AIDS in historical perspective. In Rosenberg C, ed. 1992. *Explaining Epidemics and Other Studies in the History of Medicine.* New York: Cambridge University Press.

Ruderman C, Tracey CS, Bensimon CM, Bernstein M, Hawryluck L, Shaul, RZ, Upshur RE. 2006. On pandemics and the duty to care: Whose duty? Who cares? *BMC Medical Ethics* 7:5. [Online]. Available: http://www.pubmedcentral.nih.gov/picrender.fcgi?artid=1459179&blobtype=pdf [accessed January 3, 2007].

Sencer DJ, Millar JD. 2006. Reflections on the 1976 swine flu vaccination program. *Emerging Infectious Diseases* 12(1):29-33. [Online]. Available: http://www.cdc.gov/ncidod/EID/vol12no01/pdfs/05-1007.pdf [accessed January 3, 2007].

Sokol DK. 2006. Virulent epidemics and scope of healthcare workers' duty of care. *Emerging Infectious Diseases* 12(8):1238-1241.

Sutton V. 2001. Bioterrorism preparation and response legislation—The struggle to protect states' sovereignty while preserving national security. *The Georgetown Public Policy Review* 6:93.

Sutton V. 2005. A multidisciplinary approach to an ethic of biodefense and bioterrorism. *Journal of Law, Medicine & Ethics* 33(2):310-322.

Trust for America's Health. 2005. *Ready or Not? Protecting the Public's Health from Diseases, Disasters, and Bioterrorism.* [Online]. Available: http://healthyamericans.org/reports/bioterror05/ [accessed January 3, 2007].

Upshur RE. 2002. Principles for the justification of public health intervention. *Canadian Journal of Public Health* 93(2):101-103.

WHO (World Health Organization). 2005a. *WHO Global Influenza Preparedness Plan.* [Online]. Available: http://www.who.int/csr/resources/publications/influenza/WHO_CDS_CSR_GIP_2005_5/en/index.html [accessed January 3, 2007].

WHO. 2005b. *WHO Checklist for Influenza Pandemic Preparedness Planning.* [Online]. Available: http://www.who.int/csr/resources/publications/influenza/WHO_CDS_CSR_GIP_2005_4/en/index.html [accessed January 3, 2007].

WHO (World Health Organization). 2007. *Cumulative Number of Confirmed Human Cases of Avian Influenza A/(H5N1) Reported to WHO.* [Online]. Available: http://www.who.int/csr/disease/avian_influenza/country/cases_table_2007_05_16/en/index.html [accessed May 16, 2007].

1

Learning from Pandemics Past

OVERVIEW

As David Heymann, Executive Director for Communicable Diseases at the World Health Organization (WHO), notes in the following essay, the past provides a prologue for any discussion of emerging infectious diseases, whether that discussion concerns the biological origins of a potential pandemic or its social repercussions. Thus, like the workshop, these chapters begin with a look backward. Here that look is focused on ethical issues raised in both the influenza pandemic of 1918-1920 and in more recent outbreaks of emerging infectious diseases as well as on the profound influence that these ethical issues exert on pandemic planning and on international public health law.

Reflecting on key outbreaks of emerging infectious disease over the past three decades, Heymann examines what these episodes reveal about the roles and responsibilities of health workers in a pandemic, the consequences of infectious disease to global trade, the challenge of providing equitable access to health-care resources, and the balance of individual rights versus public welfare. He describes how increasing recognition of the threat posed by emerging infectious diseases led to greater international cooperation in reporting and responding to disease outbreaks, as illustrated during the first outbreak of severe acute respiratory syndrome (SARS) and as embodied by recent revisions to the International Health Regulations.

The chapter's second paper, by medical historian Howard Markel, organizes common elements in the social experience of pandemic disease into narrative frameworks, thereby providing additional insights into legal and ethical issues in pandemic mitigation. He also describes a more specific application of historical data from the influenza pandemic of 1918-1920: evaluating the effectiveness of nonpharmaceutical interventions to reduce the transmission and impact of infectious disease. While Markel's research indicates that such efforts may have contained influenza in some U.S. communities, he acknowledges that implementing similar strategies in the future would be far from straightforward, given the increased mobility of populations, as well as the influence of civil liberties on public health policy.

Heymann's and Markel's workshop presentations were complemented by remarks from D.A. Henderson of the University of Pittsburgh Medical Center, leader of the quarter-century campaign by the World Health Organization to eradicate smallpox (Henderson, 1999). He noted that several factors made smallpox a uniquely favorable target for elimination: the virus infects only humans; it is not infectious until a rash appears; it spreads primarily through face-to-face contact; those who recover from the disease have permanent immunity; and its vaccine, which provides long-lasting protection, does not require refrigeration. Beyond these advantages, Henderson attributed the success of the smallpox eradication campaign—the first and only successful attempt to eliminate a human infectious disease from the planet—to its judicious use of available resources in host countries, its broad goals that could be achieved in multiple ways, and its support of a wide range of clinical, epidemiological, and operational research.

Henderson also explored the ethical implications of the smallpox campaign's central strategy, the vaccination of 80 percent of the world's population—which, he reported, proved a far more viable means of disease control than either quarantine or isolation. He noted that advocates of disease eradication consider immunization to be an important element of distributive justice, since the benefits of vaccination extend to all members of a community; however, eradication also raises the possibility that individual rights will be compromised if mandatory vaccination becomes necessary.

Acknowledging that top-down disease eradication programs often compete for resources with bottom-up basic health initiatives, Henderson argued that providing community-wide smallpox vaccination did serve the needs of basic health services—particularly since it provided a model for vaccinating against other important diseases. Indeed, the eradication of smallpox gave birth to an infectious-disease-management paradigm for immunization programs that, by 1990, had achieved its goal of vaccinating 80 percent of the world's children against six major diseases: tuberculosis, diphtheria, pertussis, tetanus, measles, and polio.

PAST AS PROLOGUE?

David Heymann, M.D.[1]
World Health Organization

Certainly the idea that what's past is prologue applies to any discussion of emerging infectious diseases, whether that discussion focuses on the biological origins of infectious outbreaks, or, as is the case in this workshop, on their social repercussions. In this brief summary, four key ethical issues related to emerging and reemerging infectious diseases are highlighted: the roles and responsibilities of health workers; the consequences of infectious disease to commerce among nations; the challenge of providing equitable access to health-care resources; and the balancing of individual rights versus public welfare. These four issues were very important, for example, during the first outbreak of severe acute respiratory syndrome (SARS) in 2002-2003—an event that ushered in a new era of international public health law. And they can also be expected to have relevance in any future emerging infectious disease outbreak.

Health Workers on the Front Line

The mission hospital in Yambuku, a very small community in the rainforest of the northern Democratic Republic of Congo, came to the public's attention in September 1976, when four Belgian sisters working there as nurses died of a hemorrhagic fever. Three of them died in Yambuku, while one was evacuated to Kinshasa for treatment and died there. A specimen of this fourth sister's blood was sent to the Centers for Disease Control and Prevention (CDC) in Atlanta, and it was this specimen that led to the original identification of the Ebola virus as the causative agent of this hemorrhagic fever.

One of the important facts about this outbreak is that it occurred in a hospital. It began in the maternity ward with a patient who had been at the hospital's outpatient clinic three days earlier. At that same time, another patient with a fever had been treated with an injection for what was thought to be malaria; afterwards the syringe used on that patient was rinsed with water and reused on a pregnant woman who was at the outpatient clinic on an antenatal visit. That syringe was likely the vehicle that transferred the then-unrecognized Ebola virus from one patient to another in the outpatient department and from there on to the maternity ward.

Another important feature of this outbreak is that it predominantly affected health workers and their contacts. In addition to the four Belgian sisters, the Ebola virus infected 13 African health workers plus many of their family members, most of whom died. The same situation occurred in 1995 in an Ebola outbreak in

[1]Acting Assistant Director General, Communicable Diseases.

Kikwit in the Democratic Republic of Congo. A patient admitted in early March infected two hospital staff members, a laboratory worker and a nurse, who in turn passed the infection on to family members. Later, insufficient infection control practices during a surgical procedure on one of the initial cases led to other health-care workers becoming infected.

Outbreaks that spread to the community through health workers are not limited to developing countries, however. In 1978, for example, a medical photographer at a research institution in Birmingham, England, became infected with the smallpox virus and, before dying, transmitted it to her parents. Health workers were also disproportionately affected in the 1957 H2N2 influenza pandemic, in which 52 percent of unvaccinated health workers in New York City and 32 percent of unvaccinated health workers in Chicago became infected themselves. The outbreak of SARS in 2003, and the risks posed to health-care workers will be discussed in detail later in this article.

The lesson is clear: Health workers and caregivers are inevitably on the front line in a pandemic. While they have an ethical obligation to provide safe care, they do so with the knowledge that they bear a high personal risk of infection.

Issues Between Governments: Infectious Disease and Commerce

Humans have long transmitted diseases over great distances. As shown in Figure 1-1, historians have traced the paths of three ancient diseases that, over the course of decades, spread across several continents. Some of these diseases are thought to have originated in Africa, others in Asia.

Today, infections emerge, reemerge, and spread around the world with such frequency that it is difficult to keep a list of them up to date. In 2000, for example, athletes participating in an international triathlon held in Malaysia contracted leptospirosis and returned to their home countries during the incubation period. While this disease is not transmitted from person to person, its presence in the athletes did create a diagnostic challenge for health-care workers around the world. More to the point, the case illustrates the potential for transmission in a world where international travel is both rapid and common. Figure 1-2 illustrates how polio spread from northern Nigeria after immunization activities were halted there in 2003. Wild type 1 poliovirus, endemic in that area, spread rapidly to neighboring countries, and thereafter—through Saudi Arabia and Yemen—as far away as Indonesia.

By the fourteenth century, governments had clearly recognized the capacity for the international spread of disease and had legislated preventive measures, such as the establishment of quarantine in Venice. In order to keep plague out, ships arriving in that city-state were not permitted to dock for 40 days. Table 1-1 briefly traces the history of surveillance and response to global disease from this

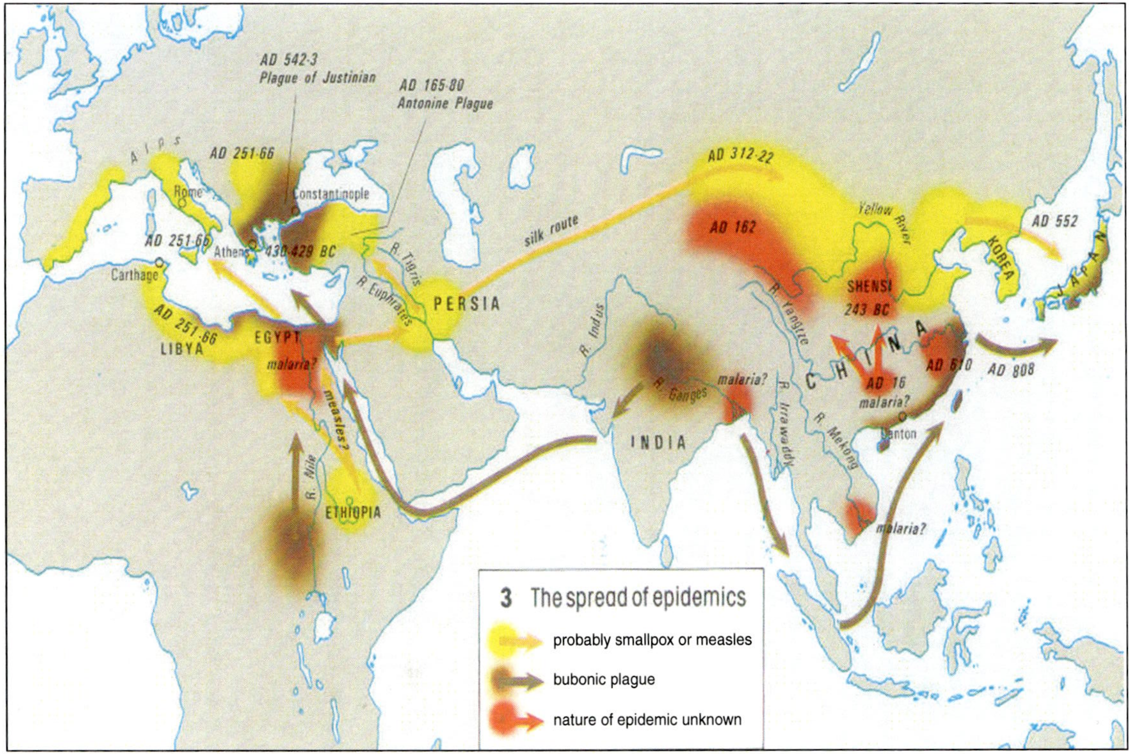

FIGURE 1-1 The spread of epidemics.
SOURCE: Heymann (2006).

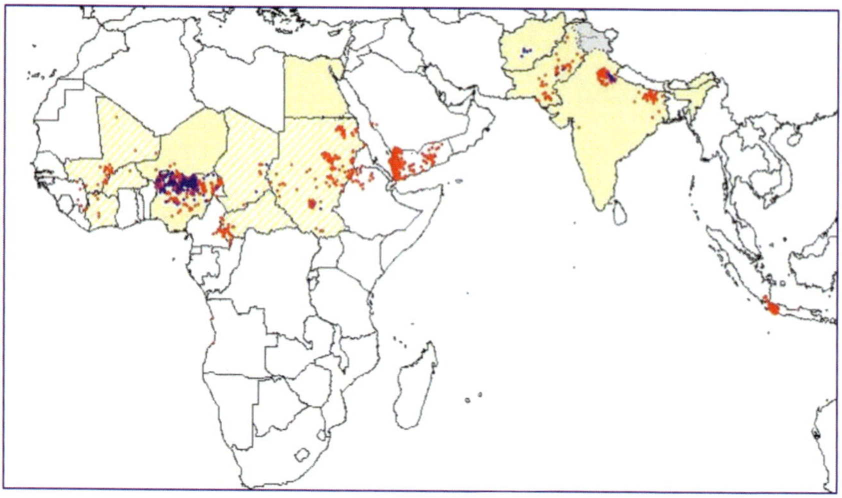

FIGURE 1-2 International spread of polio from Nigeria in 2003.
SOURCE: Heymann (2006).

TABLE 1-1 From Quarantine to International Health Regulations: A Framework for Global Health Surveillance and Response

1374	Venice	Quarantine for Plague
1851	Paris	First International Sanitary Conference
1947	Geneva	WHO Epidemiological Information Service
1951	Geneva	International Sanitary Regulations
1969	Geneva	International Health Regulations

SOURCE: Heymann (2006).

quarantine to the 1969 adoption of the International Health Regulations (IHR).[2] WHO developed these regulations, along with guides for ship sanitation and for hygiene and sanitation in aviation, as a way of minimizing the international spread of disease while interfering as little as possible in world trade, transportation, and travel.

The IHR requires that WHO be notified whenever cholera, plague, or yellow fever occur, but given today's vast number of global microbial threats, the regulations are clearly outdated. The IHR also provides guidance to ports, airports, and

[2]For more information on the evolution of the International Health Regulations see Annex 1-1, pages 59-60.

frontier posts about preventing the entry of infected travelers as well as preventing the proliferation or entry of disease vectors, such as mosquitoes and rats. The regulations specify the maximum precautionary measures that countries may adopt in order to protect themselves from the three reportable diseases as well as the measures that they should undertake to deal with infectious diseases in general. Reports of cholera, plague, or yellow fever received by WHO are published in the *Weekly Epidemiological Record.*

In 2005, a substantial revision and modernization of the IHR was adopted. The revision addresses a long-standing problem: that countries often do not report the presence of infectious diseases within their borders because they fear the economic consequences of doing so. Trade sanctions resulting from infectious disease are often more severe than necessary, as happened, for example, following the discovery that people had contracted variant Creutzfeldt-Jakob disease by eating beef from cattle in the United Kingdom (UK). These cattle had been infected with prions that caused bovine spongiform encephalopathy (BSE). Many countries reacted by banning imports from the UK, even after the UK had taken measures that probably rendered its products more secure from BSE than those of many of these same countries. The result was that the UK lost billions of dollars in trade.

The lesson, then, is that the international spread of disease—or the threat of its spread—reduces commerce with affected areas. Governments must, therefore, attempt to balance two competing goals: to prevent infectious disease from crossing their borders while simultaneously minimizing the economic impacts of disease-related restrictions on travel and trade.

Securing Equitable Access to Health-Care Resources

Some epidemics recur year after year because the affected populations do not have access to the appropriate vaccines and drugs. This was once the case with smallpox, and it is currently true of meningitis. Every dry season in Africa, meningitis causes large epidemics with high fatality rates in a belt of countries stretching from Senegal in the west to Ethiopia in the east. In 1996, during the largest recent outbreak, 250,000 people were infected and 25,000 died. Many of these deaths occurred because vaccine did not reach affected communities fast enough.

In response, a collaboration established in the late 1990s between Doctors Without Borders, the International Federation of Red Cross and Red Crescent Societies, the United Nations Children's Fund (UNICEF), and WHO attempted to address this problem by pre-purchasing and stockpiling vaccine for distribution to countries that reach a critical threshold of meningitis cases. In addition the Gates Foundation has provided support to a partnership between the Program for Appropriate Technology in Health (PATH) and WHO to develop an affordable conjugate meningitis vaccine that will be incorporated into routine immunization programs in Africa. Hopes for success are high, as a similar international partnership dealt with smallpox in much the same way, and that disease is now relegated to the history books.

Polio has presented a similar challenge. In 1988, polio was reported in 125 countries that lacked adequate access to the polio vaccine; by 2005, only four countries had not yet interrupted transmission of the virus. (Because the disease has spread internationally, however, seven countries are currently experiencing polio outbreaks.) Thanks to a partnership of Rotary International, the Centers for Disease Control and Prevention (CDC), UNICEF, WHO, and a group of international financial partners, there is now equitable access to polio vaccine for children throughout the world.

In the event of an influenza pandemic, however, access to vaccine will be extremely limited, particularly in the developing world. The global influenza vaccine manufacturing capacity is limited to the approximately 300 million doses of seasonal influenza vaccine, while the global population is 6.6 billion. These doses are produced and distributed each year, mainly within industrialized countries, in formulations that must track slight changes in this constantly mutating virus. This shortfall—the difference between a capacity of 300 million doses and a population of 6.6 billion—presents a challenge that can only by met through global preparation and action.

Public Health Measures: Balancing Individual Rights and the Common Good

During the smallpox eradication campaign, vaccines were offered to targeted populations using a ring vaccination strategy: vaccinating all households around that of the infected person and vaccinating any contacts that could be traced. In some cases, people were coerced to accept vaccination in the interest of the common good. Today, travelers through Asian airports during the influenza season receive mandatory thermal scans as they move through immigration. Passengers with fevers are taken aside, examined and, at times, prevented from traveling. These are only two of the many instances in which individual rights have been sacrificed in the interest of protecting the public from infectious diseases. Such choices represent the most common, yet most vexing, challenges in addressing microbial threats.

SARS: Revisiting the Past, Ushering in a "New World Order"

At a meeting in 1995 to establish an emerging infections program at the WHO, a panel of expert advisers decided that an updated version of the IHR could provide a valuable global framework for alert and response as well as for global communication and collaboration. WHO had previously collected information pertaining to the IHR solely from national governments, but the decision was made to risk using—and acting upon—information from existing regional and global networks as well. These included the Global Emerging Infections System (GEIS) of the U.S. Department of Defense; the Global Public Health Intelligence Network (GPHIN), which is developed and managed by the Public Health Agency of Canada; the WHO global laboratory network for influenza; and a broad array of region-specific surveillance networks, such as those sponsored by the Asia-Pacific Economic Cooperation (APEC) and the Association of Southeast Asian Nations (ASEAN). All of these were linked to construct a "network of networks" which was named the Global Outbreak Alert and Response Network (GOARN).

It was GOARN (with information derived from GPHIN and GEIS) that on November 16, 2002, reported to WHO that an outbreak of respiratory illness had occurred in Guangdong Province, China. WHO, as is its practice, went to the Chinese government in confidence. The Chinese government, which had been investigating the outbreak, found isolates of influenza B in 31 persons in the affected area. The findings were confirmed by an influenza laboratory in the WHO network, and the Chinese government decided that the outbreak was due to normal, seasonal influenza B activity. On that occasion, the alert system worked well.

Illness Among Health Workers

On February 11, 2003, GPHIN registered rumors about an outbreak of atypical pneumonia in Guangdong Province among health workers. On February 14, the Chinese government reported that 305 such cases had occurred, including five that resulted in death, but it described the outbreak as "under control." WHO remained very concerned, however, in part because the 1957 and 1969 influenza pandemics are thought to have originated in southern China and partly because the outbreak had included a large number of health workers, which suggested a possible amplification of transmission in the hospital setting. The WHO network of influenza laboratories, which looks for novel influenza viruses that might have pandemic potential, was notified of this outbreak, as were the WHO offices in countries throughout the world.

On February 19, 2003, the WHO Global Influenza Surveillance Network reported that a 33-year-old Hong Kong man and his nine-year-old son had contracted influenza A H5N1—the first time this avian virus had been detected in humans in Hong Kong since its initial appearance in 1997. The father and son had traveled through Guangdong Province to Fujian Province—where the family's 8-year-old daughter had developed a severe respiratory illness, died, and been buried—and had then returned to Hong Kong. When viewed together, these events created great concern that the Guangdong outbreak might represent the onset of an influenza pandemic.

A pandemic was indeed in its early stages, but not of influenza. Instead, a previously unknown coronavirus began to spread internationally in February

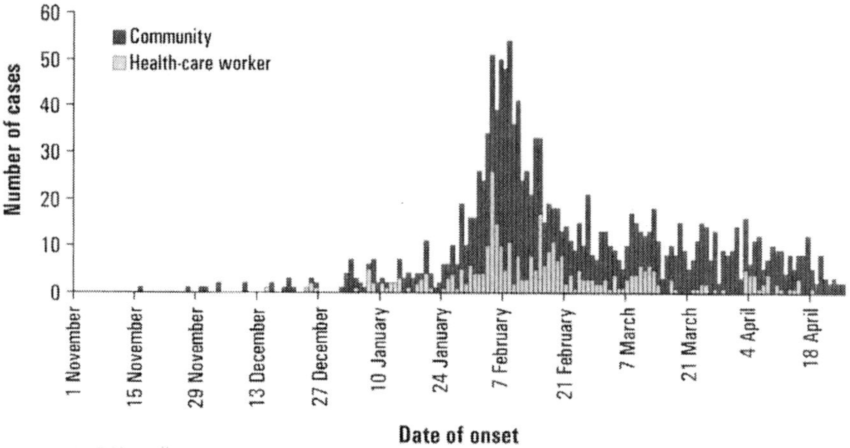

FIGURE 1-3 SARS epidemic curve, China, 2002-2003.
SOURCE: Xu et al. (2004) and Heymann (2006).

2003, when a doctor who had treated patients in the Guangdong Province traveled to a Hong Kong hotel. There, during a single day, he was somehow able to transmit the virus to other hotel guests who afterwards traveled to Canada, Singapore, and Vietnam, and to one who later entered a hospital in Hong Kong. That index case and secondary cases resulted in the infection of 219 health workers. When the chain of infection was traced backward, it was discovered that the original outbreak in China had proceeded sporadically until December 2002 when the first known hospital worker was infected (see Figure 1-3). The disease spread within the hospital and hospital workers began to amplify transmission of the virus by spreading it to their family members.

Like the Ebola outbreaks described earlier, SARS transmission was amplified and spread through the infection of health workers. And it was not only health workers who treated SARS patients who were at risk: Dr. Carlo Urbani, the WHO staff member who investigated the first SARS case in Vietnam, himself became infected and died from the disease in March 2003. Even since the SARS pandemic was contained, several minor outbreaks have occurred among researchers who were exposed to the virus in laboratory accidents.

Global Alert and Containment

On February 26, 2003, the WHO office in Hanoi reported the case of a 48-year-old businessman with high fever, atypical pneumonia, and respiratory failure who had recently traveled to China and Hong Kong. The seriously ill patient was placed on a respirator and transferred back to Hong Kong. By early March, 77 health-care workers in Hong Kong and 7 in Vietnam were reported to have atypical pneumonia, and it was clear from virological studies that the cause was not influenza. Based on this information, WHO issued its first global alert on March 12: a moderate announcement informing governments, ministries of health, and journalists that a new and highly virulent atypical pneumonia of unknown cause was occurring in Vietnam and Hong Kong.

By March 14, WHO had received reports from Canada and Singapore of persons fitting the case definition of the new atypical pneumonia. The next day, Dr. Michael Ryan, the WHO duty officer, was awakened at 2 a.m. by a call from the ministry of health in Singapore. The official reported that a doctor who had treated patients with atypical pneumonia in Singapore and who had gone to New York for a medical meeting had become ill and was on his way home on a Singapore Airlines flight. WHO worked with the government of Germany to have this patient removed from the airplane in Frankfurt and isolated there; his wife, who also was sick by that time, was also hospitalized in Frankfurt.

On March 15, the situation appeared grave: over 200 patients were infected with the new illness, which apparently was caused by an infectious agent unknown to medical science. Health workers appeared to be at greatest risk of infection. Antibiotics and antivirals were not effective against the illness, which was spreading within Asia and to Europe and North America. Clearly, this was an emerging infection, but its course was impossible to predict. It might become endemic in humans like HIV/AIDS; it might become endemic in animals; or it might pass through two or three generations and attenuate, as monkey pox had done.

Facing this uncertainty, WHO embarked upon a program of global alert and containment. It began by giving the disease a name—sudden acute respiratory syndrome—that would not stigmatize any region or country and by providing increasingly more detailed case definitions as information about the disease evolved. The health organization issued emergency guidance for travelers and urged airlines to watch for and report illness among passengers who had traveled to affected areas. And, at the same time, WHO enlisted support for investigating SARS from institutions represented by GOARN. In all, the effort would grow to involve 115 experts from 26 institutions in 17 countries. Field teams were sent to affected areas, while other experts remained in Geneva to supplement WHO staff.

The electronic networks connecting WHO with countries and regions across the globe made it possible to use real-time information to control the spread of SARS. It soon became clear that, despite the alert issued on March 12, SARS was being spread internationally by air travelers. In some instances, infected travelers were found to have spread the virus to other travelers during the flights themselves. The most famous incident occurred on a China Airlines flight from Hong Kong to Beijing on which many different persons became infected. A number of Asian businessmen who traveled to areas with outbreaks returned home apparently in good health, only to later develop SARS; meticulous epidemiological investigation strongly suggested in-flight transmission of the virus.

In response to these events, on April 2 WHO boldly made additional recommendations that exceeded the existing three-disease framework of the IHR. Airlines serving areas where local transmission of SARS was occurring were advised to actively screen departing passengers using two simple questions: Did the traveler have a history of contact with persons with SARS or with a syndrome similar to SARS? Did the traveler have a fever, cough, or other signs and symptoms? If travelers answered either question in the affirmative, WHO recommended that countries not permit those persons to depart.

The Consequences of Public Health Measures on Individual Rights and National Economies

Each country with local transmission of SARS determined how it would control the further spread of disease. In Hong Kong, information on each person with SARS, including their name and their contacts, was recorded in a police database normally used to identify clusters of crime, with the goal of identifying clusters of SARS. The names entered into the database were also provided to immigration officials in order to prevent those individuals from traveling abroad. Furthermore, Hong Kong used remote screening to detect fever in all airline passengers and required each passenger to have a health declaration. Passengers with fevers were prevented from traveling either within or outside of the country.

On May 2, WHO concluded that environmental transmission of SARS had occurred in one apartment complex in Hong Kong. (It was later determined that this incident was precipitated by several coincident factors). This discovery, along with other cases of SARS that could not be traced back to contact with an infected person, caused great concern that the virus was now spreading outside a confined setting, such as a hospital, and into the general community. This concern led WHO to advise international travelers to postpone nonessential travel to areas with SARS outbreaks. These recommendations, which were distributed on the World Wide Web, came with a major financial cost to those areas where the infection was located. Airline travel to affected areas all but halted, resulting in more than $30 billion in losses in Asia, according to estimates by the Asian Development Bank estimates.

Revision of the International Health Regulations

Within four months of beginning containment activities, and without the use of novel drugs or vaccines, all chains of human-to-human transmission were broken, the SARS virus was driven out of its new human host, and the outbreak was declared over. Several factors contributed to this success: vigorous national containment activities, including case identification, case isolation, contract tracing, surveillance, and quarantine of contacts; well-publicized international travel recommendations; and also an element of good fortune. WHO, working with its many international partners, was able to provide risk assessment and communication during the SARS pandemic that allowed countries to deal with this emerging infectious disease. Although WHO advisories were not always clearly understood by the general public, governments were alerted to the existence of the pandemic and were guided as to how to manage its risks. And, fortunately, SARS's relatively slow rate of spread allowed time for epidemiological investigation and containment. It is unlikely that an outbreak of influenza would afford such opportunities.

The global response to SARS illustrates the importance of moving beyond the passive role WHO previously played in addressing infectious threats. Indeed, the current vision for the best way to deal with emerging infectious diseases is of a world on constant alert, prepared to detect and respond to international infectious disease threats within 24 hours, using the most up-to-date means of global communication and collaboration. And the new IHR framework is a major move toward achieving this vision.

That framework can accommodate all emerging infectious diseases of international concern, including pandemic influenza. Possible outbreaks are detected using information from networks, as well as from individual countries, and, in a significant break with the past, reports other than official government notifications can be used by WHO to alert the world to an event of international concern. When an outbreak is suspected, a confidential decision-tree analysis is conducted with the affected country. If such an outbreak proves to be of international importance, WHO will support collaborative risk- and evidence-based development of public-health measures and a national containment plan.

It was the SARS outbreak, more than anything else, which led to the realization that the previous approach to emerging infectious diseases was no longer workable. SARS made it clear, for example, that WHO required a revised and greatly strengthened legal framework—the IHR—to obtain reports of infectious diseases from sources other than countries, even though such an action represents a potential infringement on national sovereignty.

With the revised IHR there is now a formal framework for proactive international surveillance and response to any epidemic that begins to spread internationally. And, in particular, the revised IHR (which will come into full legal force in June 2007) will guide responses to any future flu pandemic. At the World Health Assembly in May 2006, WHO was asked to coordinate immediate voluntary implementation of all provisions in the revised IHR relevant to the current avian influenza situation and the related threat of a pandemic. An emergency Influenza Pandemic Task Force has been established for this purpose, and it held its first meeting in Geneva on September 25, 2006. During that meeting, the experts considered the criteria for declaring the start of an influenza pandemic and asked whether the current pandemic alert should be raised to a higher level.

Given the current situation—a novel influenza virus is causing sporadic human cases but remains poorly adapted to humans—the experts decided that there is no need to alter the present level of alert. The group also examined a variety of other issues, including ways to improve the sharing of H5N1 influenza virus isolates and information on genetic sequences, the updating of diagnostic reagents and test kits, the development of a pandemic vaccine, and the monitoring for virus strains resistant to currently available antiviral drugs. Their conclusions and recommendations are included in formal report that will be reviewed by the WHO Executive Board in January 2007 and the World Health Assembly in May 2007. These activities form part of the strengthened mechanisms by which WHO and its many partners maintain vigilance for emerging microbial threats and activate defense mechanisms that protect the international community from threats to its health and shocks to societies and economies.

CONTEMPLATING PANDEMICS: THE ROLE OF HISTORICAL INQUIRY IN DEVELOPING PANDEMIC-MITIGATION STRATEGIES FOR THE TWENTY-FIRST CENTURY

Howard Markel, M.D., Ph.D.[3]
University of Michigan at Ann Arbor

Introduction

Although my initial charge for this workshop was to discuss and review the history of the 1918 influenza pandemic, I suspect that that most of you have heard or read that story many times, especially over the last few years. Instead, I propose to widen the dialogue so that we may consider the broader—and I think richer—history of epidemics and pandemics in the American experience.

Because an epidemic represents a living, social laboratory it provides a useful window through which to view the resilience and efficiency of a particular society's administrative structures, its political and social strengths and shortcomings, and its engagement with rumor, suspicion, or outright bad behavior. After all, epidemics are hardly quiet occasions; they are experienced and responded to in real time by the affected community and then later discovered, heralded, and explained by historians like me. As a result, the historical record of these events is especially rich and provocative (Briggs, 1961; Rosenberg, 1987; Rosenberg and Golden, 1992).

In what follows I will link some of the lessons learned from pandemics past to the quandaries that policymakers are grappling with today in response to a potential influenza pandemic and other microbial threats. And, given that we

[3]George E. Wantz M.D. Distinguished Professor of the History of Medicine; Professor of Pediatrics and Communicable Diseases; Director, Center for the History of Medicine.

simply do not have that much solid data on the means of mitigating or containing worst-case scenario influenza pandemics in our modern era, I will discuss why exploring the historical record of the 1918-1920 pandemic may help uncover a body of clues and suggestions. What makes that record so compelling to me as a historian of infectious diseases is that the 1918-1920 American influenza experience constitutes one of the largest databases ever assembled in the modern, post-germ-theory era on the use of non-pharmaceutical interventions to mitigate pandemic influenza in urban centers. Policy makers, on the other hand, may find it more compelling that the record allows them to have the chance to observe how large numbers of people respond when a pandemic appears but vaccines and antivirals are neither effective nor widely available. History suggests that when faced with such a crisis, many Americans—and more formally, American communities—will adopt, in some form or another, what they perceive to be effective social-distancing measures and other nonpharmaceutical interventions (NPI). This is precisely what the nation did in 1918-1920, resulting in a wide spectrum of results and outcomes. A critical question is, Can we make sense of and exploit this historical data to inform decisions today on how best to employ or discard various NPI strategies? And, if so, can we evaluate their costs and benefits in a manner that includes a polished set of social, legal, and ethical lenses?

No one can claim that history provides some magical oracle of what to expect in the future. Human history simply does not work that way. It may move in distinct and recognizable patterns, but this is quite different than repeating itself in predictive cycles. Yet despite those limitations, historians, since at least the days of Thucydides, have contributed nuanced and contextualized views of how past dilemmas emerged or evolved and have offered useful models of the resolution of those dilemmas. These views and models merit our attention.

In particular, historians have been trying for millennia to make sense of epidemics, and we can learn much from studying their conclusions. What follows are but two of the many useful models that historians have developed for analyzing the structure of epidemics.

The Four Acts Model of an Epidemic

When considering the broad scheme of an epidemic or pandemic as a social phenomenon, perhaps the best study that I know of is not a study at all but is rather the remarkable novel by Albert Camus, *The Plague*—a text I routinely assign to all my students hoping to learn anything about epidemics. Indeed, the eminent historian Charles Rosenberg uses the novel in his seminal essay "What Is an Epidemic?" to gain insights into the nature of an epidemic, combining the observations from fiction with decades of scholarship documenting three of the most serious public-health crises of human history—the devastating cholera pandemics of 1833, 1845, and 1866 (Rosenberg, 1992). From these considerations

Rosenberg characterizes the unfolding of an epidemic as a dramaturgic event, usually in four acts, with a distinct but somewhat predictable narrative plot line:

During the first act, "progressive revelation," members of a community begin to acknowledge an increasing number of cases and/or deaths resulting from the spread of a particular contagious disease. Camus's *The Plague* demonstrates this pattern with one of the most memorably disgusting opening scenes in all of literature:

> When leaving his surgery on the morning of April 16, Dr. Bernard Rieux felt something soft under his foot. It was a dead rat lying in the middle of the landing. On the spur of the moment he kicked it to one side and without giving it a further thought, continued on his way downstairs. Only when he was stepping out onto the street did it occur to him that a dead rat had no business to be on his landing. . . .

In the pages that follow Dr. Rieux finds many more dead rats along the streets of Oran, but it takes a great deal of hectoring, cajoling, lecturing, and—perhaps most critical when chasing after an epidemic—precious time to convince his fellow townspeople that there is, in fact, a serious problem threatening the entire community's health. This lethargic response is not restricted to the pages of fiction. Slow acceptance and delayed courses of action in the face of contagious threats are common features in the history of human epidemics. In some cases this tardiness is ascribed to "failure of the imagination," a reason that may be *au courant* but that is decidedly uninformative. More often the delayed acknowledgment of an epidemic can be explained by the fact that acknowledging it would threaten various interests or strongly held beliefs, from the economic and institutional to the personal and emotional.

Act two, "managing randomness," involves the society creating an intellectual framework within which the epidemic's "dismaying arbitrariness" can be understood. Readers of *The Plague* will recall the heated debate over causation of the epidemic that took place between the doctor, who subscribes to a modern, scientific approach to understanding the plague, and the Catholic priest, who preaches that the plague's visitation was an act of divine retribution for sinful lifestyles which thus demanded repentance. This dichotomy in understanding deadly disease, with religion or morality on one hand and science on the other, was a hallmark of many societies in the past, and we should not discount the role that religious, spiritual and cultural beliefs and practices can play in mitigating, containing, or inflaming an epidemic in our own era.

The third act is "negotiating public response." Once an epidemic is recognized, the public typically demands that collective action of some kind be taken. The history of epidemics is littered with tales demonstrating the importance of bold, decisive leadership and the costs of ineffective or incompetent crisis management. As many historians observing the tug of war between the public and those charged with protecting their health have noted, the operative word in public health is "public." It is generally necessary to develop a strong consensus among the multitudes constituting a community, taking into account varying cultural values and attitudes, social and class hierarchies, and economic and political imperatives, and if those efforts fail, it is often the case that little is accomplished in any attempt to rein in disease.

Act four, "subsidence and retrospection," is perhaps the most vexing phase of an epidemic, at least to those involved in public health management and epidemic-preparedness planning. Epidemics often end as ambiguously as they appear. Or, to lift a phrase from the poet T.S. Eliot, they end "not with a bang, but a whimper." Specifically, once an epidemic peters out and susceptible individuals die, recuperate, or escape, life begins to return to its normal patterns, and healthy people begin to place the epidemic in the past. Although this act can conclude with deep retrospection and action in terms of preparedness for subsequent epidemic events, more often in American—and, in fact, the world's—history, the curtain closes on a note of complacency or even outright amnesia about the event. A critical question, therefore, is how a community or government maintains credibility in its warning systems, maintains public support for costly preparedness planning, and keeps the public alert but not alarmed, panic-stricken, or completely disengaged.

This four-act model of epidemics is an excellent starting point for our contemplation of pandemics, but, of course, not all microbial threats will follow such a straightforward narrative structure. For that reason, many historians of epidemics have taken a different tack and set out to understand epidemics by identifying their major ingredients or features. This leads to a different model of the structure of epidemics and pandemics.

Major *Leitmotivs* of Pandemics

In my own work over the past 16 years I have attempted to identify and describe critical *leitmotivs* that have appeared repeatedly in epidemics and pandemics across time. To this end I have analyzed numerous pandemics, including the Black (bubonic) Plague of the Middle Ages; smallpox in the seventeenth and eighteenth centuries; the cholera pandemics of the nineteenth century (in 1832, 1845, 1866, and 1892); the influenza pandemics of 1880, 1918, 1957, and 1968; childhood infectious disease epidemics of the early twentieth century, including diphtheria, polio, and scarlet fever; small pox epidemics of the nineteenth and twentieth centuries; and also contemporary crises involving HIV/AIDS, tuberculosis, SARS, and other newly emerging infectious diseases (Markel, 1999, 2000, 2001, 2004; Stern and Markel, 2004; Markel and Stern, 1999, 2002).

Not all of the themes that I have identified in this work will appear in each epidemic or pandemic. Instead they should be viewed as major ingredients of an epidemic with the understanding that the precise mix of the themes can change from era to era and disease to disease. These *leitmotivs* include the following:

Thinking about epidemics is almost always framed and shaped—sometimes in useful ways, sometimes not—by how a given society understands a particular disease to travel and infect its victims.

People living in eras when microbes were not considered to be the cause of epidemic diseases responded to these threats differently from people living in eras when the role of microbes was understood. Well into the nineteenth century, for example, experts and lay people alike believed that many epidemics and contagious diseases were spread through polluted air—or miasma, from the Greek word for defilement of the air or pollution. The miasmatic theory of disease held that toxic emanations emerged from the soil or from rotting organic material or waste products and caused specific epidemic diseases such as cholera, typhus, and malaria. Given the foul odor that pervaded every urban center of this era, the belief that it was an unhealthy force makes a good deal of sense, but when this theory was in vogue it led to public-health approaches that were very different from those taken today. Aside from calls for quarantine, most attempts to manage an epidemic centered on cleaning up and disinfecting streets, sewers, privies, and other dirty parts of the urban environment. This trend changed markedly in the mid-to-late nineteenth century with the advent of the germ theory of disease, and it continues to be revised, refined, and fine-tuned today as we learn more and more about microbial ecology, evolution and genomics. Still, old ideas about contagion are often slow to die and, like fevers of unknown origin, have the power to recrudesce; as a result, many people today have ideas about the cause and spread of particular infectious diseases that are markedly different than the principles we teach in the medical school classroom (Duffy, 1992).

The economic devastation typically associated with epidemics can have a strong influence on the public's response to a contagious disease crisis.

An order of quarantine, which closes a port or a city to foreign travelers or goods, costs communities a great deal of money and creates great hardships for individuals. It is not surprising, then, that during the international sanitary conferences of the mid-nineteenth century, merchants were often vocal opponents of any efforts to prevent or contain disease that might have had the effect of impeding commercial enterprises and the flow of capital. Such concerns are particularly salient in today's world, given the existence of a globalized marketplace in which a rapidly growing percentage of the world's population does business, especially since the emergence of India, China, and the former Communist bloc nations.

There are two sides to this equation however. While increased global commerce can certainly contribute to the spread of a pandemic, it also sets up conditions that encourage more effective responses to a pandemic. Epidemics cost the business community a lot of money, and, in particular, the cost of a human-to-human avian influenza pandemic would be, according to all reliable projections, simply staggering. The threat of such losses could therefore encourage developing nations faced with a brewing epidemic to communicate more openly with Western nations in the hope that their greater financial resources could help them rapidly contain or mitigate the outbreak (Stern and Markel, 2004).

The movements of people and goods and the speed of travel are major factors in the spread of pandemic disease.

It is no coincidence that the rise of bubonic plague pandemics during the Middle Ages (as well as the invention of the formal concept of quarantine) coincided with the advent of ocean travel and imperial conquest. As humans traveled in wider and wider circles, so too did the germs that inhabited them. During the nineteenth century, four devastating cholera pandemics were aided and abetted by the transoceanic steamship travel of millions of people. By the close of the 19th century, journeys from Europe or Asia to North America required a travel time of 7 to 21 days, which gave most infectious diseases ample incubation periods and facilitated their recognition by health officers at the point of debarkation. It is quite different today, when the main mode of international travel, commercial jet planes, allow people to travel anywhere in the world in less than a day. Indeed, a recent study in *PLoS Medicine* details how seasonal influenza can mirror peaks and valleys in air travel (Brownstein et al., 2006). Yet while the natural response to a pandemic might be to limit air travel, either by an international edict or by the natural response of people to avoid travel by commercial airliner during such a crisis, such a response would pose a new set of troubling and potentially damaging consequences.

Our fascination with the suddenly appearing microbe that kills relatively few in spectacular fashion too often trumps our approach to infectious scourges that patiently kill millions every year.

In 2003, for example, society's response to SARS—which affected approximately 8,000 people and killed 800—was much more dramatic than its response to tuberculosis, which infected 8,000,000 and killed 3,000,000 that same year. In 2001 there was a similar disproportion in the response to anthrax, which threatened only a few, and to the ongoing global pandemic of HIV/AIDS, which kills 2,000,000 people a year. An even more egregious example is the lack of widespread attention to the common scourges of lower respiratory tract infections and diarrheal diseases, which kill millions on an annual basis (Markel and Doyle, 2003; Achenbach, 2005). Unfortunately, it will be impossible to know until long after the money and resources have been committed—and perhaps only after a flu pandemic has actually occurred—whether influenza was the right microbe to focus upon instead of one of the host of other emerging and re-emerging infectious threats that we face. Perhaps the more salient question for our discussion today is how we can apply the lessons of SARS, influenza, AIDS, bioterrorism, and other microbial threats to develop a comprehensive and global plan against contagion.

Widespread media coverage of epidemics is hardly new and is an essential part of any epidemic.

The media has the power both to inform and to misinform. Because the media powerfully shapes the public's perception of an epidemic, the details of how popular communication is carried out are of utmost importance. Today's coverage of pandemic events differs from previous eras in the technology, speed, and variety with which news reports are generated. In the early twentieth century, for instance, American consumers relied heavily on an extensive print media, whereas consumers today can turn to a panoply of newspapers, magazines, television, radio, cable, Internet sites, Web logs, and discussion groups. That does not mean that Americans today are better informed. In the early twentieth century there were multiple daily editions of newspapers in every major city and large town and a great deal of superb reporting on epidemic threats, allowing a majority of Americans to be well-informed on a wide swathe of scientific issues as they were understood at the time. It is hardly a new phenomenon how physicians, public-health officials, and others simultaneously accommodate, inform, and, at times, correct the press. Nonetheless there is no question that the breadth of media genres—and the demographics of their consumers—is far greater today than in previous eras, and there is no doubt that the media has a far greater ability to provide consumers with both useful information and misinformation.

A dangerous theme of epidemics past is the concealment of the problem from the world at large.

Across time many nations or states have concealed news of an epidemic to protect economic assets and trade. In 1892, for example, the German government initially concealed—and therefore exacerbated—that year's cholera pandemic because of fears that closing the port of Hamburg, at the time the largest port in the world, would mean economic ruin for many (Markel, 1999; Evans, 2005). At other times concealment efforts have been motivated by nationalistic bias, pride, or politics, as was the case with South Africa and HIV in the 1990s, China during the first months of the SARS epidemic of 2003, and, over the past few years, Indonesia and avian influenza (IOM, 2004, 2005). Regardless of the reasons for concealment of a public-health crisis, from the political to the purely mercenary, secrecy has almost always contributed to the further spread of a pandemic and hindered public health management.

One of the saddest themes of epidemics throughout history has been the tendency to blame or scapegoat particular social groups.

History has demonstrated too often that social groups already deemed to be "undesirable" by the population at large are most at risk for harsh or inappropriate treatment in times of crisis, no matter whether the crisis is a product of infectious disease, natural disasters, or simply social unrest. At many points in American history, especially during the nineteenth and early twentieth centuries, the implicit assumption that social undesirability was somehow correlated with increased risk of contagion has led to the development of harsh policies aimed at the scapegoats rather than the containment of a particular infectious microbe.

There are many examples of scapegoating across time, such as the widespread American assumption during the cholera pandemic of 1892 that any case of cholera discovered in the United States had been brought from Eastern Europe in the bodies of impoverished Jewish immigrants, the demonization of the Chinese in the 1900 bubonic plague outbreak in San Francisco, and, more recently, the stigmatization of gay men and Haitians during the early years of the AIDS epidemic in the United States (Markel, 1999, 2004; Kraut, 1994; Grmek, 1990).[4] At many—but certainly not all—points of time, poor people have been disproportionately affected by epidemics and pandemics. Public-health policies that place blame on victims or, worse, on perceived victims can have many negative consequences, including the misdiagnosis of the healthy and isolating or quarantining them with unhealthy people; social unrest, legal entanglements,

[4]For a broader look at the history of quarantine, infectious diseases and public health, particularly as they pertain to influenza, see: Mullet CF. 1949. A century of English quarantine, 1709-1825. *Bulletin of the History of Medicine* 23(6):527-545; McDonald JC. 1951. The history of quarantine in Britain during the 19th century. *Bulletin of the History of Medicine* 25(1):22-44; Hardy A. 1993. Cholera, quarantine and the English preventive system, 1850-1895. *Medical History* 37(3):250-260; Rosen G. 1958. *A History of Public Health*. New York: MD Publications; Duffy J. 1992. *The Sanitarians: A History of American Public Health*. Chicago: University of Illinois Press; Schepin OP, Yermakov WV, eds. 1991. *International Quarantine*. Madison, CT: International Universities Press: 125-58; Risse G. 1988. Epidemics and history: ecological perspectives and social responses. In Fee E, Fox D. 1988. *AIDS: The Burdens of History*. Berkeley: University of California Press: 33-66; Winslow, CEA. 1967. *The Conquest of Epidemic Disease: A Chapter in the History of Ideas*. New York: Hafner; Crosby AW. 1989. *America's Forgotten Pandemic: The Influenza of 1918*. New York: Cambridge University Press; Hoehling AA. 1961. *The Great Epidemic*. Boston: Little Brown & Co; Kolata G. 1999. *Flu: The Story of the Great Influenza Pandemic*. New York: Touchstone Books; Barry J. 2003. *The Great Influenza*. New York: Viking. For more literary versions of the drama of epidemic disease and quarantine, see: Boccaccio G. 1931. *The Decameron*, Translated by J Payne. New York: Modern Library; Defoe D. 1948. *A Journal of the Plague Year*. New York: Modern Library; Camus A. 1948. *The Plague*. Paris: Knopf. Ibsen H. 1988. *An Enemy of the People*. Translated by J McFarlane. Oxford, UK: Oxford University Press; Lewis S. 1925. *Arrowsmith*. New York: Harcourt Brace; IOM (Institute of Medicine). 2005. *The Threat of Pandemic Influenza: Are We Ready?* Washington, DC: The National Academies Press, especially the chapters by J Taubenberger, pp. 69-89 and by L Simonsen, et al., pp. 89-114.

and infringements of civil liberties; and extremely counterproductive behaviors by those targeted as diseased. Such negative results have the potential to detract in a major way from efforts to contain or mitigate a contagious disease.

Both historical constructs of pandemics—the four-acts model and the identification of *leitmotivs*—proved helpful in our center's analysis of the 1918-1920 influenza pandemic. For example, when examining the second wave of the pandemic, which stretched from September to December 1918, Rosenberg's four act-play metaphor provides a useful framework for understanding the rise and fall of that phase of the pandemic. Ultimately, however, the Rosenberg model works best for a single-phase epidemic rather than a multiphasic pandemic such as the entire four-wave flu pandemic of 1918-1920.

The *leitmotiv* model can also be a useful lens through which to view the 1918 pandemic, but with one key exception: the social scapegoating *leitmotiv* was not all that loud. I suggest that this was because the pandemic spread so rapidly and ubiquitously among all sectors of American society (especially among those 20-45 years of age). That does not mean, however, that we should assume that this unsavory feature of epidemic disease could not rear its head in the present or future. One has only to recall the SARS epidemic of 2003 and the short-lived but well-publicized ban on all Asian exchange students at the University of California at Berkeley, to name one recent example, to realize that it can still happen here.

All of the other *leitmotivs* described above did feature prominently in the 1918 influenza pandemic. For example, during the 1918 pandemic it was very common for local business owners to oppose nonpharmaceutical interventions that seriously affected their economic health. School and business closings, restrictions on travel, and even the use of face masks often proved to be quite contentious issues. Furthermore, many warnings of an influenza pandemic in the early summer of 1918 went unheeded; indeed, the stacks of medical libraries are filled with rarely read public health reports published in the years before the flu pandemic that urged the creation of more hospital beds and isolation wards as well the development of better diseases surveillance and containment strategies (Markel, 1999). And once the flu crisis was over, little was done to rectify public health administrative problems that were exposed by the 1918-20 pandemic.

Other *leitmotivs* that played significant roles in the pandemic include how the media interpreted the contagious spread of influenza and reported on these events; the role public health risk communications played in containing or mitigating the spread; the internecine rivalries between local, state, and federal health agencies and political leaders; suppression of reporting of cases (in 1918, this was often because privately practicing physicians did not want to lose control of—and remuneration from—their paying patients by reporting and referring them over to public health departments); the unclear etiology of influenza; ineffective vaccines against the wrong organism; and, of course, issues of travel, particularly the mass movements of soldiers around the country and then to the European theater of what we now refer to as World War I.

Although historians by nature are hesitant to predict the future, I feel quite comfortable in suggesting that most or all of these themes will again be part of whatever emerging infectious disease crises we face in years to come. And while I cannot tell you what the exact proportion or precise mix of ingredients in this recipe will be, I do think history provides us with many thought-provoking, broad-brush strokes with which to think about pandemics.

The Power and Limits of Historical Inquiry

To investigate how historical inquiry can inform the planning of pandemic mitigation strategies, one must first be aware of the limits of this approach. Let us begin by describing the historian's laboratory: the archives. A good way to think about archival research is to imagine your life being recorded by a historian. Every day the scholar would file a report and store that document in a bank of file cabinets that, by the end of your life, would presumably hold many reams of paper. Imagine, then, that a fire destroys most of that room, with only occasional file folders from discrete periods of your life surviving. With few exceptions, such spotty records are what historians deal with in their inquiries, and much of our knowledge of the past depends on the supporting archival materials that were actually saved. Furthermore, some archival materials may not be entirely reliable or may simply be unavailable, and sometimes historians may misinterpret the materials, creating yet more problems. Many times, lacunae in the historical record are so great that we can only hypothesize or speculate about what may actually have occurred.

Moreover, when one studies the history of epidemic disease, a whole new set of highly specialized records becomes important. A historian needs to be intimately familiar with the relevant era's collection of epidemiological data, its medical terminology (the same term can mean different things in different medical eras), its surveillance and containment methods, and its medical and microbiological understandings of the cause and spread of the disease. For the 1918-1920 influenza pandemic there are many cases where critical numerical population and case-incidence data were not recorded or were recorded in a manner less consistent than we would demand of a prospective study conducted today. Such gaps constitute significant challenges and even roadblocks in any historical study.

One also needs to be familiar with the social, cultural, and intellectual history of the region under study and to know its differences from and similarities to our contemporary era. For example, someone studying the 1918 flu epidemic should know that the United States of that time had many similar features to the modern era: rapid transportation in the form of trains and also automobiles, although certainly many fewer automobiles than we have today; rapid means of communication in the form of telegraph and telephone; large, heterogeneous populations with substantial urban concentrations (although many more Americans lived in rural environments in 1918 as compared to the present); a news and information system that was able to circulate information on the pandemic widely; and a broad spectrum of public health agencies at various levels of government.

Conversely, there are also many striking contrasts between that era and our own. For example, the legal understanding of privacy and of civil and constitutional rights as they relate to public health and governmentally directed measures (such as mass vaccination programs or medications) has changed markedly over the past eight decades. Furthermore, public support of and trust in these measures—along with trust in the medical profession in general—has changed significantly over this time, especially with regard to vaccines and medications. This can be seen, for example, in the recent spate of lawsuits filed because of vaccine failures or because of perceptions that vaccines may have significant and dangerous side effects. Other features of the modern world that need to be considered when studying the historical record of the 1918 pandemic in order to inform contemporary policymaking include the speed and mode of travel, particularly the development of high-volume commercial aviation; immediate access to information via the Internet and personal computers; a baseline understanding among the general educated population that the etiological agents of infectious diseases are microbial; and advances in medical technology and therapeutics which have vastly changed the options available for dealing with a pandemic.

Another important aspect of American society circa 1918 that was markedly different from the present is how daily commercial transactions are carried out. In 1918 there were no supermarkets, refrigeration was primitive, and a limited variety of preserved foods were available for purchase. Consequently, consumers often needed to shop daily at multiple locations, such as grocers, produce vendors, bakeries, and butchers. Moreover, there were no credit cards, and personal checking accounts were typically employed only by the affluent, so frequent visits to banks for cash were not uncommon. Indeed, for ordinary citizens in 1918 the United States was almost entirely a cash economy. So while the closure of a bank during an epidemic in 1918 might be explained as a public health measure, for the many Americans who had lived through the Depression of 1893 as well as other boom and bust cycles, such an action might well be misconstrued as a failure of the bank itself, and, as such, it had the potential to create civil unrest. As a result, the last public spaces to close during the 1918 pandemic—after theaters, schools, churches, restaurants, and saloons—were often banks and other financial institutions.

Today, on the other hand, a number of daily functions of life can be accomplished with little or no human interaction—provided you have the economic and educational resources to carry them out. Banking and credit transactions, the ordering and delivery of food via the Internet, entertainment, and personal and business communication, to name just a few, can all be carried out by large numbers of Americans in a way that can allow them to minimize human contact and thus shield themselves somewhat from the spread of contagious disease (Germain, 1996; Chandler, 1980; Blackford, 2003; Rothbard, 2002). Nevertheless, as recent disasters have shown, many Americans have little in the way of an economic safety net, and their restricted access to financial resources and even basic needs of living could have a deleterious affect on disaster-containment strategies.

The Defense Threat Reduction Agency/Department of Defense Escape Communities Study

The overwhelming majority of histories of the 1918 influenza pandemic focus on its widespread carnage. Consequently, our research group was surprised to uncover the archival remnants of a handful of American towns or institutions that emerged from the virulent second wave of the pandemic—September to December 1918—with relatively few influenza cases and no deaths.

In July of 2005, we were asked by the Defense Threat Reduction Agency of the U.S. Department of Defense to study these "escape communities" of 1918 because the Pentagon was contemplating what to do with personnel essential to the nation's security in the event of a pandemic. The crucial question we were being asked was if the historical experiences of these escape communities might reveal some strategy to keep a small, but specific, sector of the population—the U.S. Armed Forces—completely free of influenza. The results of this year-long, in-depth archival study proved somewhat vexing.

Some of these so-called escape communities that we studied, such as the village of Fletcher, Vermont (population 737) were too small to suggest that their success resulted from anything more than remote location, the uneven attack rates of the virus, and good fortune. Others—like the Trudeau Tuberculosis Sanatorium in Saranac Lake, New York, and the Western Pennsylvania Institution for the Blind, in Pittsburgh—were already *de facto* quarantine islands because of the era's prevailing views toward the confinement of the contagious and the disabled.

Two communities, the U.S. Naval base at Yerba Buena Island, one mile from the busy port of San Francisco, and the mining town of Gunnison, Colorado not only escaped the pandemic, they also had carried out a particularly extensive menu of restrictive public health measures (i.e., nonpharmaceutical interventions). Under the bold, decisive direction of astute public health officers, the still-healthy island and mountain towns essentially cut off all contact with the outside world to shield themselves from the incursion of influenza before it arrived in their vicinity, a measure we termed protective sequestration. In a nation besieged by flu, Yerba Buena and Gunnison boasted zero mortality and almost no cases of infection over a lengthy time period.

When planning for pandemics, it is tempting to focus on the apparent success of protective sequestration at Yerba Buena and Gunnison. But lest we be too eager to adopt such measures widely today, we must recall that one of these communities was literally an island directed by the bold, iron hand of a naval commander who could isolate his men from flu-ridden San Francisco. The other was a small, homogeneous, and well-run mining town situated high in the Rockies that could barricade its roads and regulate its railways.

Historical analysis of the few communities around the world that did manage to escape the 1918 influenza pandemic (including Australia and American Samoa) reveals an obvious but admittedly not terribly practical prescription: live in a remote area, preferably an island or mountain community, that can wall itself off from human contact. On the other hand, there are tantalizing suggestions that all these escape communities experienced much milder third waves of the pandemic when compared to neighboring communities.[5]

The CDC/Michigan Historical Study of Nonpharmaceutical Interventions Taken by 43 U.S. Cities During the Second and Third Waves of the 1918-1920 Pandemic

Beginning in August 2006 the Center for the History of Medicine at the University of Michigan Medical School, collaborating with the CDC's Division of Global Migration and Quarantine, embarked upon a study of the non-pharmaceutical interventions taken by the 43 most-populated cities in the continental United States (population > 100,000) in the second and third waves of the 1918-1920 influenza pandemic.

During the 1918 pandemic, a broad menu of NPI was executed in different American cities that have captured our attention including making influenza a reportable disease; isolation of the ill; quarantine of suspect cases and families of the ill; closing schools; protective sequestration measures; closing worship services; closing entertainment venues and other public areas; staggered work schedules; face-mask recommendations or laws; reducing or shutting down public transportation services; restrictions on funerals, parties, and weddings; restrictions on door-to-door sales; curfews and business closures; social-distancing strategies for those encountering others during the crisis; public-health education

[5]For the full report of this study, see: Markel H, Stern AM, Navarro JA, Michalsen J. 2005. *A Historical Assessment of Nonpharmaceutical Disease Containment Strategies Employed by Selected U.S. Communities during the Second Wave of the 1918-1920 Influenza Pandemic*. Defense Threat Reduction Agency: U.S. Department of Defense. [Online] Available: http://www.med.umich.edu/medschool/chm/influenza/assets/dtra_final_influenza_report.pdf [accessed December 28, 2006]. To consult all of the primary source materials that comprised this report, see: The University of Michigan Center for the History of Medicine. The 1918-1920 Influenza Pandemic Escape Community Digital Document Archive. [Online] Available: http://www.med.umich.edu/medschool/chm/influenza/index.htm [accessed December, 28, 2006]. For the abbreviated published report of this study, see: Markel H, Stern AM, Navarro JA, Michalsen JR, Monto AS, DiGiovanni Jr C. 2006. Nonpharmaceutical influenza mitigation strategies, U.S. communities, 1918–1920 pandemic. *Emerging Infectious Diseases* 12(12): 1961-1964. [Online]. Available: http://www.cdc.gov/ncidod/EID/vol12no12/pdfs/06-0506.pdf [accessed May 1, 2007].

measures; and declarations of public health emergencies. The motive, of course, was to help mitigate community transmission of influenza.

Over the next twelve months we will endeavor an historical epidemiological analysis of the application of NPIs in these communities during 1918-1919 with the goal of informing the potential use of NPIs in future pandemics. At present, no rigrous, systematic historical and epidemiological study exists on the relationship, positive or negative, between influenza case incidence and death rates during the 1918 pandemic and the NPIs taken at different points of time by the most-populated urban centers in the United States. Our principal aim is to fill this intriguing and pertinent lacuna.

Working with a team of epidemiologists, historians, and statisticians, based both at Michigan and the CDC, we are now engaged in the rather arduous task of digging up every municipal report from the 43 large cities in the continental United States during the 1918-1920 pandemic—many of which reside in dusty unmarked boxes or storage units of libraries that have rarely (if ever) been consulted in the secondary historical literature on the pandemic. Further, we will analyze a wide body of U.S. census data, including weekly mortality reports from this period as well as 86 different daily newspapers produced over an 8-month period, records from U.S. military bases, hospitals, and universities, and a huge number of other historical documents and papers from libraries and archives across the nation. When completed, the final report and its supplementary Web-based influenza archive will constitute a widely accessible version of the largest single collection of nonpharmaceutical intervention data taken in the United States during the 1918-1920 influenza pandemic.

Every detail, whether it is the number of the dead in a particular city for a particular week or the political battles being reported in the press, will be compared with at least two other sources for verification. Similarly, in each of the cities studied we will consult at least two newspapers that have been identified in terms of political party affiliation, editorial policy, and circulation figures.

As Alfred Crosby has noted in his classic book, *America's Forgotten Pandemic: The Influenza of 1918*, in human terms the pandemic was not one overarching story but instead "thousands of separate stories" with different origins and outcomes for the influenza victims, their families, and their communities (Crosby, 1989). We do not promise any oracular commandments for pandemic preparedness, but we are confident that our fine-grained, rigorous, and scholarly historical epidemiological analysis of these American cities will significantly inform those who are considering the application, utility, policies, and design of nonpharmaceutical interventions today.

Conclusion

When contemplating pandemics it is clear that precise shapes and contours of the next influenza pandemic will be strikingly different from those of the past.

But there is a positive side to this change over time. Specifically, this is essentially the first pandemic in human history where we will have had some semblance of advance warning—and hence, the opportunity to prepare. Similarly, with the advances in virology, surveillance, rapid communications, modern computing, and epidemic modeling, there is the exciting hope that we can apply all these methods to a pandemic's rapid mitigation, if not containment or outright prevention. As such, I am historically optimistic that lessons from both the past and present can help us devise effective and also ethically and socially appropriate strategies to mitigate the microbial threats that inevitably loom on our horizon.

REFERENCES

Achenbach J. 2005 (December 7). Can we stop the next killer flu? *Washington Post* W10.

Blackford MG. 2003. *A History of Small Business in America.* Chapel Hill: University of North Carolina Press.

Briggs A. 1961. Cholera and society in the nineteenth century. *Past and Present* 19(1):76-98.

Brownstein JS, CJ Wolfe, Mandl KD. 2006. Empirical evidence for the effect of airline travel on the inter-regional influenza spread in the United States. *PLoS Medicine* 3(10):e40.

Chandler AD. 1980. *The Visible Hand: The Managerial Revolution in American Business.* Cambridge, MA: Belknap Press.

Crosby AW. 1989. *America's Forgotten Pandemic: The Influenza of 1918.* Cambridge, UK: Cambridge University Press: 66.

Duffy J. 1992. *The Sanitarians: A History of American Public Health.* Urbana: University of Illinois Press.

Evans RJ. 2005. *Death in Hamburg: Society and Politics in the Cholera Years, 1830-1910.* New York: Penguin Books.

Germain RN. 1996. *Dollars Through the Doors: A Pre-1930 History of Bank Marketing in America.* Westport, CT: Greenwood Press.

Grmek MD, Ed. 1990. *History of AIDS: Emergence and Origin of a Modern Pandemic.* Translated by RC Maulitz and J Duffin. Princeton, NJ: Princeton University Press.

Henderson DA. 1999. Eradication: Lessons from the past. *Morbidity and Mortality Weekly Report* 48(SU01):16-22.

Heymann D. 2006 (September 19). *Emerging Infectious Diseases: Past Is Prologue.* Keynote address at the IOM's Forum on Microbial Threats public workshop entitled "Ethical and Legal Considerations in Mitigating Pandemic Disease," Washington, DC.

IOM (Institute of Medicine). 2004. *Learning from SARS: Preparing for the Next Disease Outbreak.* Washington, DC: The National Academies Press.

IOM. 2005. *The Threat of Pandemic Influenza.* Washington, DC: The National Academies Press.

Kraut AM. 1994. *Silent Travelers: Germs, Genes, and the "Immigrant Menace."* New York: Basic Books.

Markel H. 1999. *Quarantine!: East European Jewish Immigrants and the New York City Epidemics of 1892.* Baltimore, MD: Johns Hopkins University Press. Pp. 85-134.

Markel H. 2000. For the welfare of children: The origins of the relationship between U.S. public health workers and pediatricians. *American Journal of Public Health* 90(6):893-899.

Markel H. 2001. Journal of the plague years: Documenting the history of the AIDS epidemic in the United States. *American Journal of Public Health* 91(7):1025-1028.

Markel H. 2004. *When Germs Travel: Six Major Epidemics That Have Invaded America Since 1900 and the Fears They Have Unleashed.* New York: Pantheon Books.

Markel H. 2006 (August 9). Bird flu: Major threat or Chicken Little? *Medscape Public Health and Prevention.* [Online]. Available: http://www.medscape.com/viewarticle/541875 [accessed December 28, 2006].

Markel H, Doyle S. 2003 (April 30). The epidemic scorecard. *The New York Times* A31.

Markel H, Stern AM. 1999. Which face? Whose nation?: Immigration, public health, and the construction of disease at America's ports and borders, 1891-1928. *American Behavioral Scientist* 42(9):1314-1331.

Markel H, Stern AM. 2002. The foreignness of germs: The persistent association of immigrants and disease in American society. *Milbank Quarterly* 80(4):757-788.

Rosenberg CE. 1987. *The Cholera Years: The United States in 1833, 1845, and 1866.* Chicago, IL: University of Chicago Press.

Rosenberg C. 1992. What is an epidemic? AIDS in historical perspective. In: Rosenberg C, ed. *Explaining Epidemics and Other Studies in the History of Medicine.* New York: Cambridge University Press. Pp. 278-292.

Rosenberrg CE, Golden J, eds. 1992. *Framing Disease: Studies in Cultural History.* New Brunswick, NJ: Rutgers University Press.

Rothbard MN. 2002. *A History of Money and Banking in the United States: The Colonial Era to World War II.* Auburn, AL: Ludwig von Mises Institute.

Stern AM, Markel H. 2004. International efforts to control infectious diseases, 1851 to the present. *Journal of the American Medical Association* 292(12):1474-1479.

Xu R-H, He J-F, Evans MR, Peng GW, Field HE, Yu D-W, Lee C-K, Luo H-M, Lin W-S, Lin P, Li L-H, Liang W-J, Lin J-Y, Schnur A. 2004. Epidemiologic clues to SARS origin in China. *Emerging Infectious Diseases* 10(6):1030-1037. [Online]. Available: http://www.cdc.gov/ncidod/EID/vol10no6/pdfs/03-0852.pdf [accessed March 26, 2007].

WHO (World Health Organization). 2007. *History of WHO and International Cooperation in Public Health.* [Online]. Available: http://www.who.or.jp/GENERAL/history_wkc.html [accessed April 12, 2007].

ANNEX 1-1 History of World Health Organization (WHO) and International Cooperation in Public Health

1830	Cholera overruns Europe
1851	First International Sanitary Conference is held in Paris to produce an international sanitary convention, but fails.
1882	International Sanitary Convention, restricted to cholera, is adopted.
1897	Another international convention dealing with preventive measures against plague is adopted.
1902	International Sanitary Bureau, later re-named Pan American Sanitary Bureau, and then Pan American Sanitary Organization, is set up in Washington DC. This is the forerunner of today's Pan American Health Organization (PAHO), which also serves as WHO's Regional Office for the Americas.
1907	L'Office International d'Hygiène Publique (OIHP) is established in Paris, with a permanent secretariat and a permanent committee of senior public health officials of Member Governments.
1919	League of Nations is created and is charged, among other tasks, with taking steps in matters of international concern for the prevention and control of disease. The Health Organization of the League of Nations is set up in Geneva, in parallel with the OIHP.

continued

ANNEX 1-1 Continued

1926	International Sanitary Convention is revised to include provisions against smallpox and typhus.
1935	International Sanitary Convention for aerial navigation comes into force.
1938	Last International Sanitary Conference held in Paris. Conseil Sanitaire, Maritime et Quarantinaire at Alexandria is handed over to Egypt. (The WHO Regional Office for the Eastern Mediterranean is its lineal descendant).
1945	United Nations Conference on International Organization in San Francisco unanimously approves a proposal by Brazil and China to establish a new, autonomous, international health organization.
1946	International Health Conference in New York approves the Constitution of the WHO.
1947	WHO Interim Commission organizes assistance to Egypt to combat cholera epidemic.
1948	WHO Constitution comes into force on 7 April (now marked as World Health Day each year), when the 26th of the 61 Member States who signed it ratified its signature. Later, the First World Health Assembly is held in Geneva with delegations from 53 Governments that by then were Members.
1951	Text of new International Sanitary Regulations adopted by the Fourth World Health Assembly, replacing the previous International Sanitary Conventions.
1969	These are renamed the International Health Regulations, excluding louse-bourne typhus and relapsing fever, and leaving only cholera, plague, smallpox and yellow fever.
1973	Report from the Executive Board concludes that there is widespread dissatisfaction with health services. Radical changes are needed. The Twenty-sixth World Health Assembly decides that WHO should collaborate with, rather than assist, its Member States in developing practical guidelines for national health-care systems.
1974	WHO launches an Expanded Programme on Immunization to protect children from poliomyelitis, measles, diphtheria, whooping cough, tetanus and tuberculosis.
1977	Thirtieth World Health Assembly sets as target: that the level of health to be attained by the turn of the century should be that which will permit all people to lead a socially and economically productive life: Health for All by the Year 2000.
1978	Joint WHO/UNICEF (United Nations Children's Fund) International Conference in Alma-Ata, USSR, adopts a Declaration on Primary Health care as the key to attaining the goal of Health for All by the Year 2000.
1979	United Nations General Assembly, as well as the Thirty-second World Health Assembly, reaffirms that health is a powerful lever for socioeconomic development and peace.
1979	A Global Commission certifies the worldwide eradication of smallpox, the last known natural case having occurred in 1977.
1981	Global Strategy for Health for All by the Year 2000 is adopted, and is endorsed by the United Nations General Assembly, which urges other international organizations concerned to collaborate with WHO.
1987	United Nations General Assembly expresses concern over the spread of the AIDS pandemic. The Global Programme on AIDS is launched within WHO.
1988	Fortieth Anniversary of WHO is celebrated. Forty-first World Health Assembly resolves that poliomyelitis will be eradicated by the year 2000.
1993	Children's Vaccine Initiative launched with UNICEF, UNDP, World Bank, and the Rockefeller Foundation.
1996	WHO Centre for Health Development opened in Kobe, Japan.
1998	50th Anniversary of the Signing of the WHO Constitution.

SOURCE: WHO (2007).

2

Planning for Pandemic Influenza

OVERVIEW

Turning from the past to the worrisome present, workshop participants representing the U.S. Department of Health and Human Services (HHS) and the Pan American Health Organization (PAHO) discussed current efforts to prepare for a potential influenza pandemic. Both of these agencies have provided considerable leadership in assessing the possible effects of pandemic influenza on individuals, communities, health-care systems, and economies.

In the United States, just as the potential consequences of pandemic influenza are many and various, so are the plans being made by the U.S. government to address these contingencies, said speaker Bruce Gellin, director of the National Vaccine Program Office of HHS. In order to illustrate both the breadth and depth of national pandemic planning, Gellin discussed the overall national strategy, the role of the HHS within that framework, and several specific initiatives undertaken by the National Vaccine Program Office to fulfill that role.

In his workshop presentation, Gellin described the National Strategy for Pandemic Influenza (NSPI) created by the Homeland Security Council (HSC) as well as the detailed NSPI implementation plan released in May 2006 (HSC, 2005, 2006). The NSPI provides an integrated framework for national planning efforts across all levels of government and in all sectors of society outside of government. The integrated response should be based on the following principles, Gellin said:

- The federal government will use all instruments of national power to address the pandemic threat.

- States and communities should have credible pandemic preparedness plans to respond to an outbreak within their jurisdictions.
- The private sector should play an integral role in preparedness before a pandemic begins and should be part of the national response.
- Individual citizens should be prepared for an influenza pandemic and be educated about individual responsibility to limit the spread of infection if they or their family members become ill.
- Global partnerships will be leveraged to address the pandemic threat.

The NSPI implementation plan (see Figure 2-1), organized in stages that correspond to the pandemic phases in the World Health Organization (WHO) global framework for pandemic influenza (WHO, 2005b), provides a detailed prescription for how the government should plan and respond to a pandemic, Gellin said. The NSPI assigns responsibilities to various government agencies and departments in the areas of international efforts, transportation and borders, human and animal health, public safety, and government continuity. The implementation plan also outlines expectations for state and local governments, the private sector, and groups and individuals deemed critical to the nation's infrastructure.

The core of the NSPI implementation plan is the specification of more than 300 actions to be taken by federal departments and agencies, Gellin explained. For each such item, the plan identifies lead and supporting agencies, outcome measures, and timelines for action. Within the area of human health, these actions include the enhancement of domestic and international disease surveillance, the procurement and distribution of countermeasures, the acceleration of research and development of vaccines, drugs, and diagnostics, and the development of international cooperation, capacity, and preparedness.

In order to demonstrate how this scheme translates into specific actions by government departments and agencies, Gellin focused on HHS and one of its areas of responsibility: pandemic vaccine development programs. He noted that HHS planning for pandemic vaccine production is governed by several assumptions. First, it is assumed that the entire global manufacturing capacity for influenza vaccine, currently estimated to be approximately 300 million doses per year, would be devoted to the production of vaccine against a pandemic strain. Second, the first trial of pre-pandemic H5N1 vaccine is assumed to require two doses per person at 90 micrograms per dose—as compared with the seasonal influenza vaccine, which requires 15 micrograms per dose—although it is possible that still-unproven antigen-sparing strategies could reduce this dosage. Third, it is assumed that the U.S. cannot rely on other countries to supply vaccine in a pandemic; the nation must therefore depend upon its sole domestic vaccine manufacturer, Sanofi-Pasteur, to supply all of its pandemic vaccine. If a pandemic commenced today and these assumptions proved correct, Sanofi-Pasteur could produce enough vaccine to immunize approximately 15 million people, or about five percent of the U.S. population. In 2005, in response to these calculations,

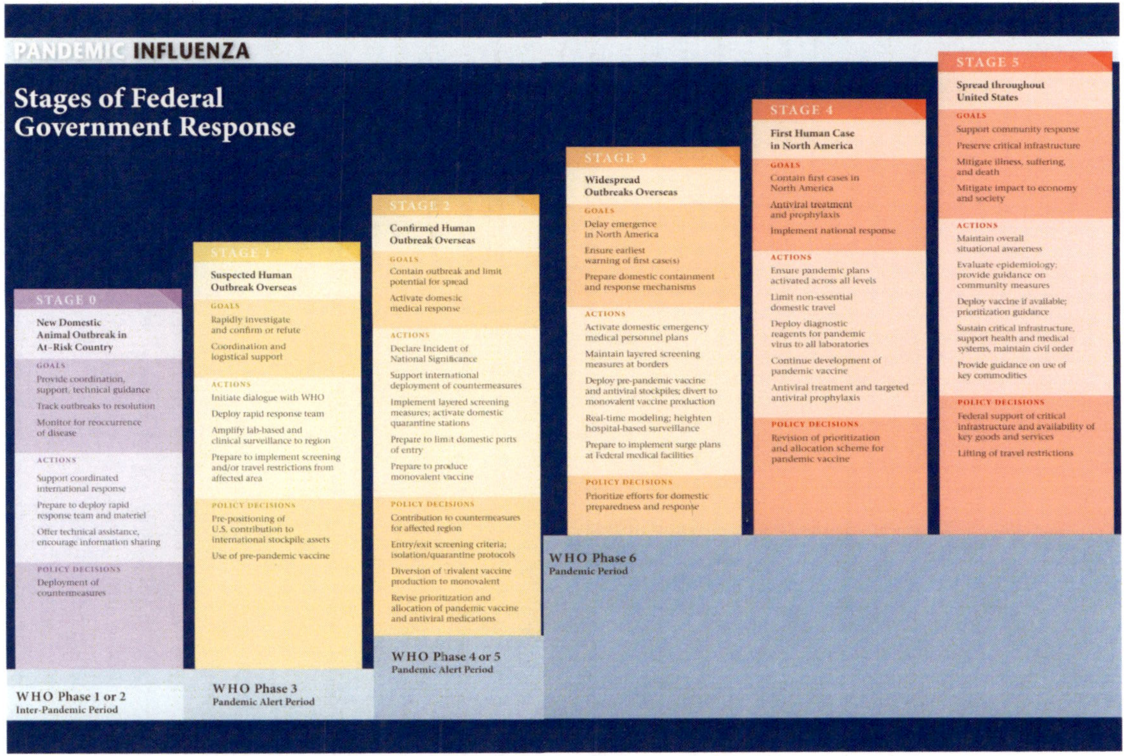

FIGURE 2-1 Stages of federal government response.
SOURCE: Gellin (2006).

the President requested $7 billion to increase domestic influenza vaccine manufacturing capacity.

One way to increase the domestic production of pandemic vaccine is to develop cell-culture influenza vaccine technology. Gellin noted that the federal government is funding an effort to license a cell-culture influenza vaccine in the United States for seasonal use, with the ultimate goal of developing domestic cell-culture facilities. While it will not hasten strain identification or speed the initial stages of vaccine development, cell-culture technology will be able to produce a far greater volume of vaccine once production begins. Moreover, Gellin said, cell-culture technology is only one of a broad spectrum of strategies being advanced by the HHS pandemic vaccine influenza program. Other goals of the agency include ensuring egg security, building pre-pandemic vaccine stockpiles, increasing vaccine production capacity, employing dose-sparing technologies, and developing broad-spectrum vaccines that could be used indefinitely against evolving strains of H5N1 influenza or other future pandemic influenza viruses.

As noted in the Summary and Assessment, workshop participants engaged in extended discussions concerning priorities for influenza vaccination, opportunities for boosting domestic vaccine production capacity, and the use of antiviral drugs for either prophylaxis or as a secondary measure when vaccines are in short supply. Gellin noted HHS Secretary Leavitt's May 2006 announcement that stocks of Tamiflu® would be pre-deployed to Asia as a first defense against a possible pandemic; the U.S. will control deployment of the stockpile and, if foreign containment efforts are not feasible, plans to return it to this country.

Reflecting on the process of developing a national pandemic strategy, workshop participants applauded the efforts that have been made to date but voiced concerns regarding the translation of national policies into local action. Steven Bice, an infectious disease specialist at Battelle Science and Technology International, noted that all responses to pandemic disease are local and advised planners that they should therefore consult with those public health officials "who will have the fight on their shoulders" in a pandemic. Rather than taking part in "a lot of top-down planning and not a lot of listening up," Bice advised the federal government to "start listening very, very carefully to the states and the locals." If they do listen, he said, they will hear that state and local officials would like guidance from the federal government on pandemic planning, including such things as feedback on the results of tabletop exercises or drills to test state and local preparedness. Conversely, he noted, the state and local perspective is lacking in the evaluation of such exercises at the federal level. In addition, Bice said, pandemic planning has been further undermined by a lack of coordination at the federal level, resulting, for instance, in turf wars between the Department of Homeland Security and HHS.

While it is widely acknowledged that an infectious disease pandemic is likely to overwhelm the U.S. medical system, workshop participants noted that the federal government has given scant attention—and even less money—to addressing this situation. D.A. Henderson projected that a pandemic would force hospitals to handle an additional 30 to 40 percent more patients than normal, a "Katrina scenario" that demands emergency planning by communities and the hospitals that serve them. The emergency plans will need to include provisions for such things as triage, the credentialing of non-physicians to provide care in an emergency, and relief from liability under such circumstances. "There is a great gaping gap here," said Henderson, who criticized government planners for focusing on the stockpiling and delivery of countermeasures of questionable efficacy rather than concentrating their efforts on "a problem which we know we are going to have."

Workshop participants also expressed concern that the U.S. pandemic influenza strategy fails to recognize our nation's dependence on and interdependence with fast-moving global markets. As noted in the Summary and Assessment, Gellin responded that the federal government's efforts to promote pandemic planning with and by the private sector—a proxy for the global economy—have begun to address this issue.

In addition to describing preparations for pandemic influenza in the United States, the session also examined such preparations in Latin America and the Caribbean under the aegis of PAHO. In their contribution to this chapter, presenter Oscar Mujica and colleagues discuss how PAHO, along with partners that include the U.S. Agency for International Development, the Centers for Disease Control and Prevention (CDC), the World Bank, the Food and Agriculture Organization of the United Nations (FAO), and the United Nations Children's Fund, has worked to encourage pandemic influenza planning in the region. In particular, they describe a key element in this process, a series of planning workshops and self-assessment exercises conducted by PAHO for national representatives of its member states and guided by the World Health Organization (WHO) Global Influenza Preparedness Plan (WHO, 2005b). The authors also report on the progress made by member states in developing national influenza pandemic preparedness plans and mechanisms for their implementation at both national and local levels.

PANDEMIC INFLUENZA PREPAREDNESS: REGIONAL PLANNING EFFORTS[1]

Oscar J. Mujica, M.D.[2]
Pan American Health Organization

Otavio Oliva, M.D.[2]
Pan American Health Organization

Thais dos Santos, B.Sc.[2]
Pan American Health Organization

John P. Ehrenberg, M.D.[2]
Pan American Health Organization

"Ver después no vale;
lo que vale es ver antes . . .
y estar preparados."[3]
—José Martí

In 2003 the 56th World Health Assembly and the 44th PAHO Directing Council issued resolutions urging countries to strengthen their capacity to prevent, detect, and diagnose influenza virus infection and to be prepared to respond to a pandemic situation (WHO, 2003; PAHO, 2003). To avoid the catastrophic consequences that would accompany a worldwide influenza pandemic, these contingency plans should be put in place now, during the inter-pandemic period, instead of waiting for the next one to strike.

In 2006 around 130 million people, or 23 percent of the total population, lived in rural areas of Latin America and the Caribbean (UN, 2006), most of them in direct contact with chickens and pigs. The FAO reports that poultry accounts for approximately 70 percent of the animal protein consumed in Latin America and the Caribbean (FAO, 2006). Also, the expanding poultry industry has become a major source of income and employment in these countries, contributing greatly to their urban and peri-urban development. A pandemic in the Region would be not only a public health problem but also a threat to food security and an

[1]Corresponding author: Oscar J Mujica, M.D., Epidemiologist, Communicable Diseases Unit, Health Surveillance & Disease Management, Pan American Health Organization, World Health Organization, 525 23rd St., NW, Washington, DC 20037. Telephone (202) 974-3974; Fax (202) 974-3656; E-mail: mujicaos@paho.org

[2]Pandemic Influenza Team of the Communicable Diseases Control Unit.

[3]Translation: "It is worthless to analyze after the fact; what counts is to anticipate what may happen . . . and be prepared."

economic disaster for the poorest populations in the rural areas and for entire national economies.

Considering the threat posed by a possible influenza pandemic, PAHO has been supporting its member states in preparing for a pandemic, as mandated by its governing bodies as well as by the 2005 Presidential Summit of the Americas. In 2005 the PAHO director created a multidisciplinary Task Force on Epidemic Alert and Response (the EAR task force) to advise, coordinate, and monitor all activities of the organization related to the planning and implementation of influenza pandemic preparedness and response. All activities of the EAR task force are guided by the revised International Health Regulations (IHR) adopted in May 2005 (WHO, 2005a). These regulations stipulate that countries develop and maintain core capacities to detect, assess, and control events of international public health importance. The inter-programmatic nature of the task force allows it to deal better with the complex process involved in IHR implementation and planning for an influenza pandemic, which requires highly coordinated efforts from a variety of sectors.

So far, PAHO's focus has been to assist member states in drafting National Influenza Pandemic Preparedness Plans (NIPPPs), taking into account the recommendations that the WHO Global Influenza Preparedness Plan offers for national measures before and during pandemics (WHO, 2005b). Box 2-1 summarizes key steps in the development and assessment of NIPPPs.

These multisectoral plans should take into account both human and veterinary health and be flexible enough to take into account various possible outcomes of a pandemic, depending on levels of viral pathogenicity and availability of resources. Subregional workshops have been carried out to train those charged with preparing NIPPPs in the use of modeling software. These modeling tools have been developed by the CDC to estimate the potential impact of a pandemic (Meltzer et al., 2000; Zhang et al., 2005; Praveen et al., 2006). The availability of such estimates helps countries keep their national plans flexible by providing them with a variety of contingencies to plan for, including a worst-case scenario where there are neither available vaccines nor antiviral medications. Table 2-1 shows a summary of estimates of potential pandemic impact in Latin America and the Caribbean for 1968-like and 1918-like scenarios which were prepared by country teams during those subregional workshops.

After member countries provided draft plans, PAHO carried out a series of self-assessment exercises where NIPPPs were evaluated using a PAHO-developed tool based on WHO's checklist for influenza pandemic preparedness planning (WHO, 2005c). The tool covers the seven core components on the WHO checklist: emergency preparedness; surveillance; case investigation and treatment; preventing spread of the disease in the community; maintaining essential services; research and evaluation; and implementation, testing and revision of the national plan. These core components are further divided into 44 main categories containing a total of 368 checkpoints for assessment.

BOX 2-1
PAHO Strategy in Supporting Member States in the Development and Assessment of National Influenza Pandemic Preparedness Plans (NIPPPs)

1. Development of draft NIPPPs
   a. Introduction of WHO guidelines for pandemic preparedness planning
   b. Introduction and application of modeling tools such as FluAid, FluSurge and FluWorkloss in order to estimate the potential impact of a pandemic
   c. Development of national action plans which include adequate development of draft NIPPPs
2. Assessment and testing of draft NIPPPs
   a. Self-assessment of NIPPPS
   b. Tabletop exercises to highlight issues of chain of command and need for multi-sectoral integration and coordination
   c. Development of action plans to address remaining gaps identified during the self-assessment and simulation exercises
3. Local implementation of NIPPPs
   a. Development of simulation drills and tabletop exercises to test local preparedness and put local contingency plans into practice
   b. Promote subnational, multisectoral training to promote the development of local contingency plans for a pandemic which adequately incorporate all pertinent areas, including surveillance, health services, disasters management, and social communication.
   c. Carrying out of tabletop exercises to test the completeness of local plans, taking into account subnational realities.
4. Monitoring and strengthening of NIPPPs
   a. Promote the use of subnational drills to assist in the monitoring of suitability of local contingency plans
   b. Promote any necessary changes in order to keep plans updated

SOURCE: PAHO (in press).

All countries in Latin America and the Caribbean participated in such self-assessment exercises. Each country delegation had participants from the areas of epidemiology, health services, laboratory diagnosis, immunization, disaster management, emergency preparedness, social communication, veterinary public health, agriculture, and international relations. The main goal of the exercises was for countries to work together in a collective, multidisciplinary, and intersectoral way to improve the preparation and implementation of their various national plans. For many of these professionals this exercise was the first time that they had sat at the same table with their peers to discuss pandemic preparedness. One of the major achievements of this interaction was the multisectoral discussion that took place about the steps required to complete the national plans of the different

TABLE 2-1 Potential Impact of a 25 Percent Clinical Attack Rate Influenza Pandemic in Latin America and the Caribbean, by Main Health Outcome and Severity Scenario, Mid-2006

	Pandemic Scenario	
Potential Health Impact	1968 Moderate	1918 Severe
Deaths	334,163 [131,630–654,960]	2,418,469 [627,367–5,401,035]
Hospitalizations	1,461,401 [459,051–1,937,503	11,798,613 [3,189,747–16,418,254]
Outpatient visits	76,187,593 [59,738,730–109,207,769]	68,470,386 [58,114,124–92,27,761]

SOURCE: PAHO (in press).

countries. In particular, the discussion highlighted the importance of joint work and integration in the contingency-planning process.

The participating countries' national plans showed varying levels of compliance with the WHO checklist, and different subregions tended to have different patterns of strengths and weaknesses in their levels of compliance.[4] In the Andean subregion, for example, national plans seemed to be most comprehensive in the areas pertaining to the management of cases, while they were not so well done in the area of research and evaluation. Caribbean countries also had difficulties with research and evaluation, but they were strong in the area of emergency preparedness. National plans for countries in Central America appear to be strongest in implementation and weakest in essential services continuity. For Southern Cone countries, development was a strength and essential services continuity a weakness. Table 2-2 presents the average compliance with WHO guidelines for each of the seven NIPPP core components in the four geographical subregions.

Emergency preparedness, the first core component in the pandemic influenza preparedness planning, is broken down into six main categories, one of which is devoted to a number of specific and fundamental legal and ethical issues as described in Box 2-2.

[4]Four sub-regions were assessed: the Andean Area, comprising Bolivia, Colombia, Ecuador, Peru, and Venezuela; the Southern Cone, comprising Argentina, Brazil, Chile, Paraguay, and Uruguay; Central America, comprising Costa Rica, El Salvador, Guatemala, Honduras, Nicaragua, and Panama (plus Cuba, Dominican Republic, and Haiti); and the Caribbean, comprising Anguilla, Antigua and Barbuda, Aruba, Barbados, Belize, Bermuda, British Virgin Islands, Cayman Islands, Dominica, Grenada, Guyana, Jamaica, Martinique, Montserrat, St. Kitts and Nevis, St. Lucia, St. Maarten, St. Vincent and The Grenadines, Suriname, The Bahamas, Trinidad and Tobago, and the Turks and Caicos Islands.

TABLE 2-2 Pandemic Preparedness Readiness in Latin America and the Caribbean. Current Compliance (percentage) with WHO Guidelines, by Core Components and PAHO Sub-Regions, Mid-2006

NIPPP core component	Andean Area	Central America	The Caribbean	Southern Cone
1 Emergency preparedness	38.6	34.6	56.7	58.5
2 Epidemiological surveillance	37.0	34.8	56.5	54.4
3 Case management	52.3	54.5	48.9	60.9
4 Population containment	20.0	38.0	37.0	64.0
5 Essential services continuity	24.5	33.3	45.2	41.9
6 Research and evaluation	10.0	40.0	15.0	30.2
7 Implementation of the national plan	40.0	60.0	30.0	50.0

SOURCE: PAHO (in press).

Among all the categories of emergency preparedness, the area of legal and ethical issues is where Latin American countries need the most work in satisfying the WHO checklist. As can be seen in Table 2-3, it is by far among the least developed areas for the countries of the Andean Area and Central America and is the second-least developed area for the other two subregions.

The workshops called attention to the need for intersectoral coordination in the development of NIPPPs. Since February 2006, when the workshops started, many countries have been engaging in intersectoral dialogue in addition to carrying out the necessary activities for the local implementation of NIPPPs. Using the momentum created by the subregional self-assessment meeting, professionals in charge of preparedness planning have strengthened their interaction with those professionals who will be responsible for implementing the plans. The discussion resulted in the roles and responsibilities for all the actors in the response becoming better defined. It was also extremely beneficial to have representatives from the legal departments of the ministries of health participate in the process, as they were able to point out amendments to health legislation that will be necessary in order to implement the NIPPP.

In addition to promoting the development of national plans, PAHO is helping its member states strengthen the mechanisms and capacities necessary to implement these plans fully. The necessary capacities include surveillance capabilities, health services, vaccine and antiviral technologies, and communication, among others. In particular, PAHO is supporting its member states in making national influenza preparedness plans operational at the local level, since National Influenza Preparedness Plans are only as effective as their local contingency plans. One important goal will be to strengthen the core competencies of member states and communities to respond to any public health emergency, as identified through the new IHR.

BOX 2-2
Core Legal and Ethical Issues to Be Considered and Assessed in the NIPPPs

Legal Issues

1. Evidence of a legislative framework in place for the national response plan
2. Legal dispositions for contingencies (maintenance of essential services and other crisis management measures)
3. Legal basis for travel and movement restrictions assessed
4. Legal basis for closure of educational institutions assessed
5. Legal basis for isolation and quarantine of infected persons or of persons suspected of being infected assessed
6. Legal basis for prohibition of mass gatherings assessed
7. Standing policy and legal basis for influenza vaccination of essential personnel assessed
8. Legal issues (liability, insurance, licensing) related to the mobilization of temporary workers assessed
9. Liability for unforeseen adverse events attributed to antipandemic strain vaccine and antiviral use considered
10. Legislative framework for compliance with the International Health Regulations in effect
11. Inclusion of pandemic influenza in national legislation for the prevention of occupational diseases considered

Ethical Issues

1. Ethical review on the limitation/restriction of access to scarce resources
2. Ethical review on the compulsory nature of vaccination of essential personnel
3. Ethical review on limiting personal freedom and movement, such as may occur with isolation and quarantine
4. Establishment of an ethical framework for research, especially when this involves human subjects

SOURCE: WHO (2005c).

Influenza pandemics have historically taken the world by surprise, leaving minimal time for health services to prepare for the surge in cases and deaths that characterize these events and that make them so disruptive (Glezen, 1996). The current situation is markedly different, however, as the world has been warned in advance. This advance warning has brought an unprecedented opportunity, especially in the Americas, to prepare for a pandemic and develop ways to mitigate its effects even in areas with problems of access to basic health services.

Evidence suggests that an influenza pandemic will be most intensely felt at the community level, especially among the young, the poor, and other vulner-

TABLE 2-3 Assessment of the Emergency Preparedness NIPPP Component, Including Legal and Ethical Issues, in Latin America and the Caribbean. Current Compliance (percentage) with WHO Guidelines, by Main Areas and PAHO Sub-Regions, Mid-2006

Emergency Preparedness	Andean Area	Central America	The Caribbean	Southern Cone
Political mobilization	58.3	55.6	66.7	88.9
Command and control	37.5	22.2	58.3	66.7
Risk assessment	40.0	50.0	37.5	40.0
Risk communication	38.5	41.7	62.5	75.0
Legal and ethical issues	11.1	13.3	40.0	60.0
Phased-response plan	33.3	33.3	66.7	100.0

SOURCE: PAHO (in press).

able groups (IOM, 2005). Despite the tremendous strides that have been made in increasing influenza pandemic preparedness at the national level, a significant challenge remains in bringing preparedness down to the subnational level—to the policy makers, practitioners, and concerned citizens who will be charged with actually implementing the national plans. As national strategies are put in place, PAHO's focus is shifting from planning and awareness-raising to increasing the local acceptance and adoption of these strategies and also to ensuring their effective implementation. In order to bridge the existing gap between planning and implementation, those who will be implementing the plans at the local level must be encouraged to take part in the national planning process. The local implementation of NIPPPs should be tested though simulation drills and tabletop exercises that test local preparedness and contingency plans. Furthermore, local-level contingency planning should be promoted and supported by the broader measure of improving the ability of member states and communities to respond to all types of public health emergencies and not just pandemic influenza.

The Region of the Americas is in the fortunate position of having the opportunity to get ready for an influenza pandemic before the virus is introduced in this part of the world. However, the fact that the region is as yet unaffected creates a false sense of security and causes pandemic preparedness to not seem so urgent. The result is that pandemic preparedness seems less important in comparison to many other competing priorities and thus falls short on the political agendas of many of the countries.

All preparations for a pandemic must be carried out under the framework provided by the revised IHR, which set a baseline level of core competencies that countries must have in order to detect and respond to any public health emergency of international concern. Similarly, existing structures and mechanisms, such as contingency plans for mitigation of emergencies and national disasters, should

BOX 2-3
Achievements of PAHO's Member States in the Development and Assessment of NIPPPs

- Professionals from varied sectors working together, often for the first time, on building national capacity to cope with a pandemic
- Countries are creating, analyzing, and refining their NIPPPs in an integrated and coordinated fashion
- Inclusion of pandemic influenza preparedness in the Health Agendas of the Regional Integration Systems (MercoSur, CariCom, CAN, SISCA)
- Basic bolstering of public health infrastructure, targeted to a possible influenza pandemic, but applicable to a varied array of public health emergencies
- Regional cadre of professionals trained in multiple aspects of influenza preparedness—health services delivery, surveillance, risk and social communication, and disasters and emergency management
- Professionals who are able to replicate training to associates and colleagues at the sub-national levels
- Commitment from trained professionals to continue influenza preparedness activities

SOURCE: PAHO (in press).

be used in preparations for pandemic influenza in order to avoid duplication of efforts and to maximize available resources.

PAHO, with its 105-year history of working with the countries in the Americas, has laid a solid foundation that can be put to work in preparing for a potential influenza pandemic. By building on its experience of supporting member states in Latin America and the Caribbean, PAHO can be an effective partner in helping them to develop and revise their NIPPPs as well as to consolidate their current achievements (see Box 2-3).

Current global threats, including influenza pandemic, require a concerted effort by all those capable of effective action. PAHO recognizes the paramount importance of partnerships in such an effort and is collaborating closely with several stakeholders. It will also continue to encourage multisectoral and multicountry approaches, such as those used during the planning exercises, to strengthen cooperation, surveillance, and communication.

Clearly, additional resources will be needed to reach a number of goals, such as stimulating counterpart support by the various countries, piggybacking on existing surveillance systems and expanding them to become population-based, and scaling up preparedness and rapid-response capabilities at the local level. And access to drugs, vaccines, and other supplies is still an unresolved issue.

Nonetheless, we believe the region has the potential to be self-sufficient. One final challenge will be to extend intersectoral involvement and commitment to the private sector, to nongovernmental organizations (NGOs), and to the academic sectors. Mechanisms for this to take place remain to be devised.

In conclusion, the threat of an influenza pandemic has revealed the weaknesses of some systems in the Americas, but it also has once again demonstrated the strong determination among the countries of the region to work together, to work fast, to overcome disparities, and to share information. Technical cooperation has served to strengthen public health in these countries. This constitutes an important global contribution and ultimately could save many lives.

REFERENCES

FAO (Food and Agriculture Organization of the United Nations). 2006. Helping prevent avian influenza in Latin America and the Caribbean. *FAONewsRoom*. [Online]. Available: http://www.fao.org/newsroom/en/news/2006/1000381/index.html [accessed January 11, 2007].

Gellin B. 2006 (September 19). *Session II: Domestic, Regional and International Preparedness Planning*. Institute of Medicine's Forum on Microbial Threats public workshop entitled "Ethical and Legal Considerations in Mitigating Pandemic Disease," Washington, DC.

Glezen WP. 1996. Emerging infections: Pandemic influenza. *Epidemiologic Reviews* 18(1):64-76.

HSC (Homeland Security Council). 2005 (November). *National Strategy for Pandemic Influenza*. [Online]. Available: http://www.whitehouse.gov/homeland/nspi.pdf [accessed January 3, 2007].

HSC. 2006 (May). *National Strategy for Pandemic Influenza Implementation Plan*. [Online]. Available: http://www.whitehouse.gov/homeland/nspi_implementation.pdf [accessed January 3, 2007].

IOM (Institute of Medicine). 2005. *The Threat of Pandemic Influenza: Are We Ready?* Washington, DC: The National Academies Press.

Meltzer MI, Shoemake HA, Kownaski M, Crosby R. 2000. *FluAid 2.0: Software and Manual to aid State and Local-Level Health Officials Plan, Prepare and Practice for the Next Influenza Pandemic (Beta Test Version)*. Centers for Disease Control and Prevention. [Online]. Available: http://www.cdc.gov/flu/tools/fluaid/ [accessed May 1, 2007].

PAHO (Pan American Health Organization). 2003. *Resolution CD44.R8: Influenza Pandemic: Preparation in the Hemisphere*. Forty-fourth Directing Council: Washington, DC. [Online]. Available: http://www.paho.org/english/gov/cd/cd44index-e.htm#resolutions. [accessed January 15, 2007].

PAHO. (In Press). *Health in the Americas: 2007 Edition*. Pan American Health Organization: Washington, DC.

Praveen D, Zhang X, Meltzer MI, Bridges CB. 2006. *FluWorkLoss 1.0: Software and Manual to Aid State and Local Public Health Officials Estimating the Impact of an Influenza Pandemic on Work Day Loss (Beta Test Version)*. Centers for Disease Control and Prevention. [Online]. Available: http://www.cdc.gov/flu/tools/fluworkloss/ [accessed May 1, 2007].

UN (United Nations). 2006. World Urbanization Prospects: The 2005 Revision. ESA/P/WP/200. Department of Economic and Social Affairs: New York. [Online]. Available: http://www.un.org/esa/population/publications/WUP2005/2005wup.htm. [accessed January 17, 2007].

WHO (World Health Organization). 2003. Resolution WHA56.19: *Prevention and Control of Influenza Pandemics and Annual Epidemics*. Fifty-sixth World Health Asembly, Geneva. [Online]. Available: http://www.who.int/gb/e/e_wha56.html/#Resolutions [accessed January 15, 2007].

WHO. 2005a. Resolution WHA58.3: *Revision of the International Health Regulations*. Fifty-eighth World Health Assembly: Geneva. [Online]. Available: http://www.who.int/gb/e/e_wha58.html/#Resolutions [accessed January 15, 2007].

WHO. 2005b. *Global Influenza Preparedness Plan: The Role of WHO and Recommendations for National Measures Before and During Pandemics*. WHO/CDS/CSR/ GIP/2005.5. Global Influenza Programme: Geneva. [Online]. Available: http://www.who.int/csr/resources/publications/influenza/WHO_CDS_CSR_GIP_2005_5/en/index.html [accessed January 16, 2007].

WHO. 2005c. *Checklist for Influenza Pandemic Preparedness Planning*. [Online]. Available: http://www.who.int/csr/resources/publications/influenza/WHO_CDS_CSR_GIP_2005_4/en/index.html [accessed January 16, 2007].

Zhang X, Meltzer MI, Wortley P. 2005. FluSurge 2.0: *Software and Manual to Aid State and Local Public Health Officials and Hospital Administrators in Estimating the Impact of an Influenza Pandemic on Hospital Surge Capacity (Beta Test Version)*. Centers for Disease Control and Prevention. [Online]. Available: http://www.cdc.gov/flu/tools/flusurge/ [accessed May 1, 2007].

3

Strategies for Disease Containment

OVERVIEW

Given limited supplies of vaccines, antiviral drugs, and ventilators, nonpharmaceutical interventions are likely to dominate the public health response to any pandemic, at least in the near term. The six papers that make up this chapter describe scientific approaches to maximizing the benefits of quarantine and other nonpharmaceutical strategies for containing infectious disease as well as the legal and ethical considerations that should be taken into account when adopting such strategies. The authors of the first three papers raise a variety of legal and ethical concerns associated with behavioral approaches to disease containment and mitigation that must be addressed in the course of pandemic planning, and the last three papers describe the use of computer modeling for crafting disease containment strategies.

More specifically, the chapter's first paper, by Lawrence Gostin and Benjamin Berkman of Georgetown University Law Center, presents an overview of the legal and ethical challenges that must be addressed in preparing for pandemic influenza. The authors observe that even interventions that are effective in a public health sense can have profound adverse consequences for civil liberties and economic status. They go on to identify several ethical and human rights concerns associated with behavioral interventions that would likely be used in a pandemic, and they discuss ways to minimize the social consequences of such interventions.

The next essay argues that although laws give decision makers certain powers in a pandemic, those decision makers must inevitably apply ethical tenets to decide if and how to use those powers because "law cannot anticipate the specifics of each public health emergency." Workshop panelist James LeDuc of the Centers for Disease Control and Prevention (CDC) and his co-authors present a set of ethical guidelines that should be employed in pandemic preparation and response. They also identify a range of legal issues relevant to social-distancing measures. If state and local governments are to reach an acceptable level of public health preparedness, the authors say, they must give systematic attention to the ethical and legal issues, and that preparedness should be tested, along with other public health measures, in pandemic preparation exercises.

LeDuc's fellow panelist Victoria Sutton of Texas Tech University also considered the intersection of law and ethics in public health emergencies in general and in the specific case of pandemic influenza. In particular, Sutton identified several "choke points"—particularly thorny ethical and legal issues—that present barriers to pandemic mitigation. In addition to the problem of leadership, which is addressed in the next chapter, these issues include the role of interdisciplinary and intersectoral approaches in decision-making; the tradeoffs between personal freedom and public good that are implicit in social-distancing measures; the global implications of quarantine and travel restrictions; the need for consistency among various disease-control policies; and the definition of appropriate, measurable "triggers" for when to impose each potential countermeasure.

The third paper in this chapter considers quarantine, one of the most ethically and legally complex tactics used in combating pandemic disease. In this article, Martin Cetron of CDC and Julius Landwirth of Yale University describe the modern practice of quarantine and its potential implementation as outlined in the U.S. Department of Health and Human Services (HHS) plan for containing pandemic avian influenza. Whenever the possibility of using a quarantine is discussed, they observe, decision makers confront the central dilemma arising from the contrast between public health ethics, which emphasizes collective action for the good of the community, and therapeutic medicine, with its focus on the individual. The authors identify various means to address this tension and offer examples of how ethical considerations can be incorporated into pandemic preparedness plans.

The chapter concludes with a three-part contribution by Joshua Epstein of the Brookings Institution: an informal discussion of the modeling process as it applies to infectious disease containment, followed by two publications in which such models are used to inform strategies for containing smallpox epidemics resulting from bioterrorism. Epstein and his group produce explicit models of disease, and, in the course of doing so, they examine and refine the assumptions upon which each model rests. Epstein observes that while models cannot replace human judgment, they can better inform our choices, and while they cannot eliminate uncertainty, models can identify crucial gaps in knowledge. To support these assertions, Epstein describes how his group collaborates with medical experts to produce disease scenarios and containment strategies (e.g., for smallpox) more robust than would be possible either through pure computation or through expert opinion alone.

Responding to Epstein's presentation, workshop panelist Timothy Germann, of Los Alamos National Laboratory, observed that models cannot address ethical and legal questions; instead models must be combined with ethical and legal judgments to make policy decisions. Epstein replied to that observation by pointing out the possibility that models of infectious disease containment could be shaped by legal and ethical considerations—introduced in the form of constraints—built into them, much as economic factors have been included in similar models. Moreover, he said, models sometimes provide information that can help resolve ethical dilemmas; for example, projections that reveal little difference in effectiveness between voluntary and mandatory quarantine.

PREPARING FOR PANDEMIC INFLUENZA: LEGAL AND ETHICAL CHALLENGES[1]

Lawrence O. Gostin, J.D.[2]
Georgetown University Law Center

Benjamin E. Berkman, J.D., M.P.H.[3]
Georgetown University Law Center

Introduction

Highly pathogenic influenza A (H5N1) has captured the close attention of policy makers who regard pandemic influenza as a national security threat. The virus is endemic in bird populations in Southeast Asia, with serious outbreaks also having now occurred in Africa, Europe, and the Middle East (WHO, 2006a, 2007a). Modeling[4] suggests that the infection will eventually affect the

[1]This is an expanded version of a two-part series: Gostin LO. 2006. Medical countermeasures for pandemic influenza: Ethics and the law. *Journal of the American Medical Association* 295(5): 554-556; and Gostin LO. 2006. Public health strategies for pandemic influenza: Ethics and the law. *Journal of the American Medical Association* 295(14):1700-1704. A longer version of this paper is published as Gostin LO, Berkman BE. 2007. Pandemic Influenza: Ethics, Law, and the Public's Health. *Administrative Law Review* 59(1): 121-175. Additionally, some of this article is based on the authors' work with the World Health Organization Project on Addressing Ethical Issues in Pandemic Influenza Planning. Professor Gostin and Mr. Berkman acknowledge the invaluable comments of their WHO working group, as well as the able assistance of Deborah Rubbens, L.L.M., and John Kraemer, JD Candidate, Georgetown University Law Center.

[2]Associate Dean and Linda D. and Timothy T. O'Neill Chair in Global Health Law, Georgetown University Law Center; Professor of Public Health, the Johns Hopkins University; Faculty Director, O'Neill Institute for National and Global Health Law at the Georgetown University Law Center; Director, Center for Law and the Public's Health, a Collaborating Center of the World Health Organization and Centers for Disease Control and Prevention.

[3]Sloan Fellow in Biosecurity Law and Policy, Center for Law and the Public's Health.

[4]A good overview of the state of current influenza containment modeling can be found at Institute of Medicine. 2006. *Modeling Community Containment for Pandemic Influenza*. Washington, DC: The National Academies Press.

entire globe through transmission mechanisms involving both birds and humans (Longini et al., 2005). The majority of avian outbreaks in Southeast Asia have been attributed to the movement of poultry and poultry products (Chen et al., 2006; Rosenthal, 2006). Similarly, international trade and travel will play a major role in transmission in human outbreaks, and frequent and widespread travel will make it difficult to contain any pandemic in humans. Even if trade and travel are severely restricted in order to limit human transmission, migratory birds will likely spread the disease by infecting birds on other continents (Normile, 2006).

So far, however, the spread of the H5N1 strain has been confined mainly to animal populations. The virus is highly contagious among birds, and also highly pathogenic (Garrett, 2005), but because of a significant species barrier, the virus is still rare in humans (WHO, 2005b). The first confirmed cases of human infection were reported in 1997. As of May 16, 2007, 306 cases of the current wave of Influenza A (H5N1) have been reported, with 185 deaths (WHO, 2007b). Most cases are attributable to close contact with infected poultry or contaminated surfaces—e.g., poultry farms, markets, backyard pets, and cock-fighting venues (Thorson et al., 2006). A few cases of human-to-human transmission have occurred, principally involving intimate household contact, but the virus is of very limited transmission competence (WHO, 2006b). The virus appears highly pathogenic, with a reported death rate exceeding 50 percent (Wong and Yuen, 2006). However, because of the possibility of under-reporting, the exact prevalence, transmissibility, and fatality rates of H5N1 remain uncertain.

Recent evidence that the 1918 pandemic was caused by an avian influenza virus lends credibility to the theory that the current strain could develop pandemic potential (Taubenberger et al., 2005; Tumpey et al., 2005). Historically, the number of deaths during a pandemic has varied greatly, depending on the number of people who become infected, the virulence of the virus, and the effectiveness of preventive measures (WHO, 2005c). Accurate predictions of mortality are thus difficult to establish, and estimates differ considerably. A mild pandemic, comparable to those in 1957 and 1968, is likely to cause the deaths of from 89,000 to 207,000 people in the United States (Garrett, 2005; Global Security, 2006) and 2 million to 7.4 million people globally (WHO, 2005d). One study that extrapolates from the severe 1918 pandemic finds that, in the absence of intervention, an influenza pandemic could lead to 1.9 million deaths in the United States and 180 million to 369 million deaths globally (Osterholm, 2005).[5] A different study, also based on 1918 data, concludes that an estimated 62 million people will die globally, with 96 percent of these deaths occurring in the developing world (Murray et al., 2006). An influenza pandemic would also result in massive economic disruption. At present, the principal economic effects are being experi-

[5]Notably, seasonal (interpandemic) influenza causes worldwide yearly epidemics resulting in 1 to 1.5 million infections.

enced in the rural areas of Southeast Asian countries and are fairly limited. They are mostly related to losses of poultry and to governmental control measures such as the culling of birds. Economic losses would become much higher if sustained human-to-human transmissions develops.

The two principal strategies for containing serious human outbreaks of influenza are therapeutic countermeasures (e.g., vaccines and antiviral medications) and public health interventions (e.g., infection control, social separation, and quarantine). Many of the barriers to effective interventions are technical and have been thoroughly discussed. This article focuses on the formidable legal and ethical challenges, which have yet to receive sufficient attention (Kotalik, 2005; Torda, 2006; Thomson et al., 2006; Kayman and Oblorh-Odjidja, 2006).[6]

Medical Countermeasures: Vaccines and Neuraminidase Inhibitors

Industrialized countries place great emphasis on scientific solutions. Vaccination and, to a lesser extent, antiviral medication (in particular, the neuraminidase inhibitors oseltamivir [Tamiflu®] and zanamivir [Relenza®]) are perhaps the most important medical interventions for reducing morbidity and mortality associated with influenza (Germann et al., 2006; Iton, 2006; Stohr and Esveld, 2004). There is also recent evidence from primate models that the 1918 H1N1 influenza strain, unlike contemporary strains, can cause an exuberant immune response, which suggests that immunity suppressants might be another means of combating at least some strains of the virus (Kobasa et al., 2007). The United States plans to devote over 90 percent of pandemic influenza spending to medical countermeasures (U.S. Congressional Budget Office, 2006; Spotswood, 2005).

Despite the promise of medical countermeasures, their use has been limited by a chronic mismatch between public health needs and private-sector control of production. Vaccine production, for example, has been unreliable even for seasonal influenza. The best way to ensure pandemic preparedness is to increase the baseline level of seasonal countermeasures. The World Health Organization (WHO) concluded that better use of vaccines for seasonal epidemics could help ensure that manufacturing capacity meets demand in a future pandemic (World Health Assembly, 2005; Gronvall and Borio, 2006). But even though this approach is good for the long-term, more immediate solutions are needed. Moreover, supply is difficult to increase because intellectual property concerns, regulatory hurdles, a lack of market incentives, limited production capacity, and fear of liability all act to curb entry into the market.

---

[6]For an example of this lack of attention to law and ethics, see Department of Health and Human Services. 2006. *Medical Offices and Clinics Pandemic Influenza Planning Checklist.* [Online]. Available: http://www.pandemicflu.gov/plan/medical.html#3 [accessed January 30, 2007]. This document purports to be a "checklist to help medical offices and ambulatory clinics assess and improve their preparedness for responding to pandemic influenza." However, it does not address the myriad legal and ethical issues that will arise.

Even if these supply problems can be overcome, it is unlikely that sufficient medical countermeasures will be available to halt the spread of a pandemic. In particular, there will likely be a significant delay in the production of a vaccine. With current technology it will take at least 6 months from the onset of an outbreak, and possibly longer, for the first doses of vaccine to be available. Furthermore, there is no guarantee that medical countermeasures will be efficacious. Experimental H5N1 vaccines may not be effective against a novel human subtype, and the pathogen may become resistant to neuraminidase inhibitors.

Public Health Countermeasures

Given the limits of medical countermeasures, a broad range of public health would likely be employed against an influenza pandemic, from relatively innocuous techniques, such as disease surveillance and hygienic measures, to considerably more restrictive interventions, such as social distancing, travel restrictions, quarantine, and case isolation. There are reasons to believe that all of these will be effective to at least some degree (Markel et al., 2006), but evidence supporting their effectiveness is scarce (IOM, 2006). The hope is that public health interventions, while incapable of completely stopping the transmission of the virus, will be able to slow the pandemic. By reducing the rate of spread of the disease, public health countermeasures can buy time for the development of medical countermeasures while also helping to ensure that the health-care system does not become overwhelmed by a surge of patients (Cetron, 2006). Unfortunately, each type of public health intervention raises serious ethical and human rights concerns.

Public Health Surveillance

Surveillance is the backbone of public health, providing the data necessary to understand an epidemic threat and to inform the public, provide early warning, describe transmission characteristics and incidence and prevalence, and assist a targeted response. Surveillance strategies include rapid diagnosis, screening, reporting, case management reporting, contact investigations, and the monitoring of trends.

It is clear that surveillance is necessary to quickly identify and respond to a pandemic influenza outbreak. The revised International Health Regulations (IHR) require member states to notify WHO of all events which may constitute a "public health emergency of international concern." Consequently, once a country identifies a signal suggesting human-to-human transmission, the country is expected to begin investigations immediately and simultaneously to notify WHO of the event. Surveillance thus comprises a crucial element of the early response to a forming pandemic.

But because governments must collect sensitive health information from patients, travellers, migrants, and other vulnerable populations, surveillance also poses privacy risks (Bayer and Fairchild, 2002). The IHR requires states to keep data "confidential and processed anonymously as required by national law," but in a crisis it can sometimes be necessary to disclose certain information without any undue delay. In such a situation, when the immediate use of the information is necessary for an important public health purpose, disclosure can be warranted, but the identity of the affected person should be protected as much as possible. A breach of the right to privacy can result not only in economic harms, such us unemployment or loss of insurance or housing, but also in social and psychological harms. For that reason, if information is released outside of the public health system, it is particularly important to avoid the inclusion of any uniquely identifiable characteristics, such as names, government identification numbers, fingerprints, or phone numbers. Cases should stay anonymous or encrypted when reasonably feasible. In every situation the rights to privacy and personal autonomy require that only the minimum amount of information necessary to achieve the goal should be released and to as few people as possible. Dignity and respect for the person should be protected.

Screening and testing can pose serious threats to a person's privacy and bodily integrity. Ideally, public health officials should receive an individual's informed consent before performing any medical tests, and education programs can help convince many people to agree to voluntary testing, but there may be rare times when mandatory testing is necessary to advance the public good. In such cases, interference with the right to bodily integrity and with the right to refuse testing may be permissible when the mandatory testing policy is clearly necessary and effective in protecting the public health, when it is performed by competent public health officials, and when the least intrusive means are used. At a minimum, compulsory testing should be limited to individuals known or at least suspected to be infected and should be done in a fair and nondiscriminatory way. The people whose privacy and autonomy are being infringed should be informed of the reasons for the infringement. And in all cases compulsion should be the last resort and used only if voluntary or less restrictive means are ineffective.

Community Hygiene and Hospital Infection Control

Hygienic measures to prevent the spread of respiratory infections are broadly accepted and have been widely used in both influenza pandemics (APHA, 1918) and also, although with uncertain benefits, the SARS outbreaks (WHO, 2003; CDC, 2005a). These hygienic methods include hand-washing, disinfection, the use of personal protective equipment (PPE) such as masks, gloves, gowns, and eye protection, and respiratory hygiene, such as the use of proper etiquette for coughs, sneezes, and spitting.

It is important that the public be informed of the need for hygienic measures, and that accurate information, including the uncertainty of the effectiveness of the recommended interventions, be provided. In past epidemics misinformation has been rampant, and this has led to substantial public anxiety, to reliance on word of mouth for knowledge, and to the purchase of ineffective and expensive products (Rosling and Rosling, 2003). The situation raises issues of distributive justice because ineffective or inaccurate communications have the greatest effects on marginalized members of society, as they are the least likely to have access to alternative credible sources of information and are the people for whom wasting resources would have the greatest adverse effects (Gostin and Powers, 2006). Furthermore, a consideration for personal dignity implies that individuals should be provided with adequate information to make informed decisions about their own health. Public education campaigns should be grounded in the science of risk communication, as the acceptability of health measures is vital to community adherence. The information disseminated through public education campaigns should be accurate, clear, uncomplicated, not sensationalistic or alarmist, and as reassuring as possible (SARS Commission, 2006).[7]

Decreased Social Mixing/Increased Social Distance

Past experience shows that one consistent response to epidemics has been to decrease social mixing and increase social distance by such means as community restrictions and voluntary social separation (WHO, 2005d; Stern and Markel, 2004). There is some limited evidence that school closings do reduce seasonal influenza transmission (Heymann et al., 2004), and it is assumed—although but not proven—that other limits on social mixing also slow the spread of respiratory disease (World Health Organization Writing Group, 2006). Thus societies faced with pandemics have often closed public places (schools,[8] childcare, workplaces, mass transit) and cancelled public events (sports, arts, conferences). As fear rises, the public itself may shun public gatherings. Predicting the effect of policies to increase social distance is difficult, as infected persons and their contacts may be displaced into other settings, and individuals may voluntarily separate in response to perceived risk. For these reasons, additional research needs to be conducted on behavior during epidemics and the effects of social distancing on transmission.

Social separation, particularly for long durations, can cause loneliness and emotional detachment, disrupt social and economic life, and infringe individual rights. Community restrictions raise profound questions about the government's right to interfere in such areas as faith (by, for instance, limiting religious gatherings), family (with, for example, restrictions on funeral attendance), and pro-

---

[7]The Canadian SARS Commission has evaluated crisis communication during that public health emergency.

[8]A review of state law authorizing school closure can be found at Hodge JG. 2006 (December 11). *Assessing Legal Preparedness for School Closure in Response to Pandemic Flu or Other Public Health Emergencies*. [Online]. Available: http://www.newfluwiki2.com/upload/Hodge.ppt [accessed January 30, 2007].

tection of the vulnerable (e.g., by making it more difficult to visit vulnerable individuals and provide them with food, water, clothing, or medical care).

Undoubtedly, most judicial systems would uphold reasonable community restrictions, but legal and logistical questions loom: Who has the power and under what criteria to order closings, and for what period of time? What threshold of disease should trigger closings, and should thresholds be different for different entities? Under what circumstances should compensation for closings be paid? What should the penalties be for non-compliance? Such questions about enforcement and the assurance of population safety are critically important, but for the most part they have not been answered.

One fear is that governments might put into effect restrictions on personal liberties that are unnecessary—implementing restrictions before they are needed, extending them past the end of the crisis, or enacting restrictions that do nothing to decrease influenza transmission. In such situations, closings would not meet the appropriate standards for either necessity or proportionality. Furthermore, it is important to remember that the cost of restrictive policies will be borne most heavily by those with the fewest resources, so errant social-distancing actions have distributive-justice implications. A final worry is that governments might use social distancing in a discriminatory fashion, scapegoating ethnic or religious minorities, or that governments might use social distancing as a pretext to crack down on dissidents who assemble to protest.

Ideally, questions of government authority and accountability should be answered by policy decisions made in an open and transparent process that encourages input from all portions of society and that is carried out before a pandemic hits. Governments should explicitly define who has the power to order social distancing strategies and for what period of time. Governments should also clearly state the criteria under which such power is exercisable and delineate the legitimate bases for any differential treatment. Penalties should be proportional to offenses and not based on irrational fears or discriminatory beliefs.

On the other hand, one must recognize that detailed pandemic influenza preparations will often not be the highest priorities for countries dealing with important and more immediate concerns. Furthermore, some countries lack the legal and governmental infrastructures to implement such an ideal plan as is outlined above. In such countries, completely determining issues of government authority and accountability prior to a pandemic may be extremely difficult. One should also note that pandemics are difficult to predict, and information acquired as a pandemic evolves may render some of what was previously believed about various social-distancing strategies obsolete.

At the very least, though, governments should dedicate themselves to non-discrimination and transparency before an influenza pandemic occurs. It is important that governments implement social-distancing policies fairly and with as broad involvement in planning as possible. This will not only make it more likely that the appropriate ethical considerations have been taken into account, but it will also improve the likelihood that the public will accept social distancing as a means to slow disease transmission. And, since compliance with social-distancing instructions will be difficult to enforce, public acceptance will be critical to such a measure's success.

Workplace and School Closings

Workplace and school closings present particularly difficult ethical issues. Apart from the uncertainty of their effectiveness, the most important issues center on the subject of distributive justice. Workplaces are vital to the livelihoods of both employers and employees, so closing them can cause severe financial hardships. In extreme cases, lost profits caused by closings may push companies to go out of business, leading to job losses and other economic hardships. Even for people who have an economic safety net, these problems can have a significant effect, but for people living at a subsistence level the effect of lost income can be far worse. If workplaces stay closed for a significant amount of time, such people may be unable to pay for shelter, food, or medicine. Similar issues are raised by school closings, which may require parents to stay at home in order to care for young children.

Ideally, public health authorities should work cooperatively with businesses, schools, and communities prior to an emergency in an effort to establish mutually agreeable closure procedures. Though governments should retain the legal power to enforce closings if absolutely necessary, it would be preferable to subsidize lost profits and incomes as necessary in order to create incentives for complying with closure requests. The latter approach was used extensively in countries affected by SARS for people placed in quarantine (Rothstein et al., 2003).

Practical constraints may sometimes make this approach impossible. The governments of many countries have more pressing needs than addressing a potential pandemic. Furthermore, some countries may be financially unable to provide compensation for closure. In 1918 each of the waves of the pandemic lasted for several months, and most locations were hit by multiple waves (Johnson and Mueller, 2002). The amount of resources needed to compensate for lost income or profits for this amount of time will be out of the reach of many of the world's governments.

In light of these constraints, governments should at least make a serious effort to weigh the risks to health and welfare from workplace closings and other social-distancing measures against those risks of disease transmission that the closings might mitigate. In different locations the balance of risks may be resolved differently, depending on resources and the number of people living at or below a subsistence level. Countries should consider tactical closures if necessary, such as closing only those entities that most facilitate transmission. For example, schools have been identified as a primary driver of seasonal influenza (Germann et al., 2006), and some believe that closing schools will slow the spread of a pandemic. Countries might also consider using closings as a means to buy time for other preparations; closings could be put in place until the level of disease in a community exceeds a predetermined level and then relaxed, with the hope of slowing the initial spread of disease through the community.

Provision of Necessities

If people are instructed to avoid public places, such as markets, stores, and pharmacies, or if those places are required to close, there will be a need for people to procure food, medicine, and other necessities in some other way. Similarly, shutting down mass transit may prevent people from being able to get to those facilities that do remain open, and it could prevent some people from being able to seek medical care. Such a situation also raises distributive-justice concerns since those people with the least resources will be least likely to be able to procure additional resources before closings occur.

Ideally governments would set up networks for the distribution of necessary provisions to citizens' homes, with a particular focus on those most in need. Such distribution should be consistent and reliable, and it should provide necessities such as food and medicine for the duration of social-distancing measures. It should also be conducted in such a manner as to minimize interaction with potentially infectious people, and those people responsible for distributing provisions should use infection-control precautions to decrease the likelihood that they will spread disease. Transportation for medical care should be provided as needed by personnel who are apprised of the risks involved in transporting potentially infectious people; these personnel should be provided with protective equipment that will allow them to guard themselves from the disease and to avoid spreading it to others. Similarly, a program should be put in place for the removal of bodies from homes in a safe and efficient manner.

Resource constraints and logistical difficulties are likely to impede such a program in many areas. Many governments may lack the resources to provide food, medicine, and other necessities to its citizens during a pandemic. Even if the resources are available, the workforce needed to conduct distribution may be absent, especially at the height of a pandemic when a substantial number of people would be ill. Furthermore, there may not be enough people willing to interact closely with potentially infectious people to allow such a system to function. Shortages of personnel may be especially likely for medical transport and mortuary services.

At the very least, governments should do what they can to facilitate the provision of resources before an area is hit by disease. To the extent possible governments should give advance warning of disease and make recommendations about what food, medicine, and other supplies should be stockpiled and in what quantities. If they are able, governments should provide such necessities for people unable to afford them on their own. Governments should provide access to medical care to the greatest extent possible, perhaps by reassigning public safety officers to this purpose. Governments should also provide a means by which people who have recovered from influenza—and who thus presumably would be immune—could volunteer to assist others in the provision of necessities.

International Travel and Border Controls

Transnational public health law has become increasingly important in global health, as evidenced by the WHO's International Health Regulations and by national health agencies' proposed communicable disease regulations (HHS, 2005a). These legal initiatives reflect WHO recommendations for border controls (WHO, 2004; IOM, 2005). Transnational containment measures can be far-reaching: entry or exit screening, reporting, health-alert notices, collection and dissemination of passenger information, travel advisories or restrictions, and physical examination or management of sick or exposed individuals. These kinds of powers were exercised in Asia and North America during the SARS outbreaks, although their effectiveness has not been established (Bell and WHO Working Group, 2004; St. John et al., 2005). The IHR also authorizes a variety of sanitary measures at borders and on conveyances, including inspection, fumigation, disinfection, pest extermination, and destruction of infected or contaminated animals or goods (HHS, 2005b). Although border protection is permitted, it can severely disrupt travel, trade, and tourism, and it should be balanced against the global economic impact.

Controls placed on international travel can also infringe upon civil liberties. The freedom of movement is a basic right protected by national laws and international treaties, but it is subject to limits when necessary for the public's health (*Shapiro v. Thompson*, 1999). In particular, some of these limits can present serious risks to privacy. For example, containment measures may require the travel industry to collect and disclose passenger data (CDC, 2005b). Such infringements on privacy rights can be justified only if there is a genuine need to obtain high-quality surveillance data and if the infringements are carried out in accordance with the fair information practices described in the surveillance section above. To avoid discrimination and to ensure proportionality, public health officials should inform the affected individuals about the reasons for the infringement, the intended use of the information, and the extent to which third parties can have access to the data.

Given the transboundary nature of travel advisories as well as the economic impact they can have on affected countries, it should be left to the WHO to issue transparent and clearly justified travel recommendations in accordance with the revised IHR. It is the responsibility of individual countries to communicate all relevant information on the emergence of a public health threat to the international community. This responsibility is related to the surveillance duties and to the issues that accompany them. Ultimately, it is the responsibility of the national government to use whatever policy instruments it has available to ensure that it can comply with the requirements of the new IHR.

Isolation and Quarantine

Isolation and quarantine are two of the oldest disease-control methods in existence and would likely be used in at least some instances during an influenza pandemic. While the terms "quarantine," "isolation," and "compulsory hospitalization" are often used interchangeably, they are, in fact, distinct. The modern definition of quarantine is the restriction of the activities of asymptomatic persons who have been exposed to a communicable disease, during or immediately prior to the period of communicability, to prevent disease transmission (Reich, 2003). In contrast, isolation is the separation, for the period of communicability, of known infected persons in such places and under such conditions as to prevent or limit the transmission of the infectious agent (Benenson, 1995). Quarantine and isolation can be accomplished by various means, including confining people to their own homes, restricting travel out of an affected area, and keeping people at a designated facility (Global Security, 2005). Whatever techniques are used, it is important to treat symptomatic, potentially infected, and non-exposed populations differently. For example, it would be inappropriate to place infected individuals in the same room as those who are only possibly exposed.

Quarantine and isolation are the most complex and controversial public health powers. Given that they involve a significant deprivation of an individual's liberty in the name of public health, quarantine and isolation expose the tension between the interests of society in protecting the health of its citizens and the civil liberties of individuals, such as privacy, non-discrimination, freedom of movement, and freedom from arbitrary detention. Although these civil liberties are protected by both universal and regional human rights declarations and conventions, large-scale public health threats can require extraordinary measures by the government. Coercive public health powers such as quarantine and isolation can be legitimately justified if the public health interests of society are carefully balanced against the freedom of the individual (Gostin, 2007). To pass the balancing test, the benefits to the public should outweigh the burdens or harms that a quarantine may place on individuals. In addition each country should comply with the Siracusa principles, a set of internationally agreed-upon legal principles that establish the conditions under which restrictions on civil liberties are justified (United Nations Economic and Social Council, 1985). These principles hold that restrictions of liberty should be legal, proportionate, necessary, and accomplished by the least restrictive means that are reasonably available.

Measures as coercive as quarantine and isolation should only be used when a disease is known through extensive scientific study to be contagious and should be limited to people who have in fact been exposed to the disease. In cases of scientific uncertainty, however, resource and time restraints can make it necessary for the government to take action without performing medical testing on each individual. There may be situations, for example, in which the availability of accurate tests and competent medical staff is limited. But at a minimum the state's power should be exercised fairly and never as a subterfuge for discrimination. In a crisis situation, reasonable suspicion based on known contact with the pandemic virus can be sufficient to issue a quarantine or isolation order. However, to ensure the legitimacy of the measures taken, the decision to use restrictive measures must be made in an open, fair, and legitimate manner. The public has a right to know the legitimate public health reasons for restricting liberty. Public health authorities should fully and honestly disclose their reasons for action and allow community participation. Such transparency will enhance public trust and the acceptance of the proposed containment measures (Markovits, 2005; Heyman, 2005).

Quarantine and isolation should be voluntary whenever possible, and, when that is impossible, they should be enforced by the least intrusive means available. Research in the aftermath of SARS showed that people understood and accepted the need for restrictive measures. Many perceived it as their civic duty and were willing to sacrifice their right to freedom of movement (University of Toronto Joint Centre for Bioethics Pandemic Influenza Working Group, 2005). However, if governments expect full voluntary compliance, the decisions need to be made in an open and fair manner, and society should ensure that those who are quarantined or isolated receive adequate care and do not suffer unfair economic burdens.

When the protection of a community's health requires that individual liberty and autonomy be restricted, the principle of reciprocity obliges society to provide those affected with the necessities of life. During quarantine, these necessities would include being housed in safe, humane conditions and receiving high-quality medical care and psychological support. Recent studies have confirmed that quarantine imposes some serious financial and psychological hardships on the affected individuals. About 30 percent of individuals quarantined for SARS, for example, suffered from posttraumatic stress disorder and depression (Hawryluck, 2004).

Distributive justice requires that officials limit the extent to which the personal and economic burdens of a public health threat fall unfairly upon individual citizens. A lack of resources and amenities should be addressed in the most fair and equitable possible way. Governments as well as national and international organizations should stockpile medical supplies and food. A pandemic influenza will require solidarity among nations and collaborative approaches that set aside traditional values of self-interest and territoriality.

Conclusion

Preparing for an influenza pandemic forces society to face a number of difficult challenges, many of which transcend the issue of mere scientific effectiveness. Public health emergencies raise serious ethical issues which are central to society's commitment to freedom and social justice. Even when effective, public health interventions can have serious adverse consequences on economic and civil liberties. It is vital that individual rights are only sacrificed when absolutely necessary to protect the public's health. As such, laws must clearly establish the criteria under which governments can exercise emergency powers. These laws must also provide adequate due process and ensure that any infringements on individual rights are minimized.

The threat of an influenza pandemic is real. If the threat manifests itself, millions of lives will be lost. Such widespread death would be catastrophic, but the tragedy would be even worse if society ignores the ethical issues discussed above. It is crucial that society decide as soon as possible how it wants to respond to these ethical concerns so that in the event of a pandemic we are equipped—scientifically as well as ethically—to deal with its impact.

ETHICAL AND LEGAL CONSIDERATIONS IN PREPARING FOR PANDEMIC INFLUENZA[9]

James W. LeDuc, Ph.D.[10]
Centers for Disease Control and Prevention

Drue H. Barrett, Ph.D.[11]
Centers for Disease Control and Prevention

Anthony D. Moulton, Ph.D.[12]
Centers for Disease Control and Prevention

Richard A. Goodman, M.D., J.D., M.P.H.[13]
Centers for Disease Control and Prevention

Kathy Kinlaw, M.Div.[14]
Emory University

Robert J. Levine, M.D.[15]
Yale University

[9]The findings and conclusions in this report are those of the authors and do not necessarily represent the views of the Centers for Disease Control and Prevention.

[10]Currently affiliated with the Institute for Human Infections and Immunity at the University of Texas Medical Branch, Galveston.

[11]Office of the Chief Science Officer.

[12]Public Health Law Program.

[13]Public Health Law Program.

[14]Center for Ethics.

[15]Institution for Social and Policy Studies, Interdisciplinary Center for Bioethics.

Introduction

A pandemic of influenza will be a global challenge that affects all sectors of society and places virtually every individual at risk, independent of social or economic status, ethnic origin, or gender. International travel will rapidly introduce novel influenza strains around the world. Major urban centers will be affected first, but soon thereafter the movement of residents will spread the disease to all but the most isolated communities. High infection rates may cause disruption of critical services across all sectors of society, and health-care facilities may be overwhelmed with seriously ill patients. Unless health-care workers are selectively protected, they may suffer disproportionately, which could lead to critical shortages of skilled health-care professionals and the further erosion of clinical care capacity. Antiviral drugs and protective vaccines will likely be in limited supply, necessitating difficult decisions on how to distribute these critical resources. Nonpharmaceutical interventions (NPIs)—including, for example, isolation of those infected, quarantine of those exposed, and other social-distancing measures—will be among the few options available to public health officials to limit the spread of infection and protect the largest number of individuals.

Preparation for pandemic influenza is a dynamic undertaking which involves many partners and different sectors of society. There must be careful preparation and coordination among all stakeholders, undertaken in a fully transparent and inclusive manner, in order to ensure equitable distribution and optimal benefit from limited supplies of antiviral drugs and vaccines. Further, successful implementation of social-distancing measures and other NPIs will require systematic attention to the important legal and ethical issues that their use raises.

Laws give public health agencies the power to act to protect the public's health through the exercise of such measures as isolation and quarantine, but laws necessarily must leave room for discretion by decision makers because it is not possible to anticipate the specifics of each public health emergency. For this reason public health officials must apply ethical reasoning on matters for which the law does not provide precise guidance. In particular, such ethical considerations should inform officials' deliberations when making difficult choices that directly affect the health and well-being of the populations they serve. In the pages that follow, we suggest ethical guidelines that will be relevant to efforts to prepare for and respond to a pandemic, discuss some of the legal issues that require consideration, and conclude with some general comments relevant to national and international preparedness efforts. It is important to recognize that scientific knowledge about effective pandemic influenza interventions is evolving rapidly. The issues and suggestions presented here, while intended to have broad applicability, nonetheless may need to be reassessed as scientific knowledge advances, as the results of various public health interventions become known, and as the pathogens under consideration themselves evolve over time.

Ethical Guidelines

In an attempt to establish systematic ethical guidelines to guide decision makers in preparing for and responding to pandemic influenza, the CDC asked the Ethics Subcommittee of the Advisory Committee to the Director, CDC, to prepare a set of guidelines for use in the prioritization and distribution of vaccines and antiviral drugs and also in the development of any interventions that might limit individual freedom and create social distancing.[16] Table 3-1 outlines the general ethical considerations that the Ethics Subcommittee describes in its document. In preparing these guidelines, the subcommittee took into account the need to have ethical perspectives provide practical direction as well as the importance of having any proposed guidelines fully vetted by those involved in planning and responding to pandemic influenza.

It is clear that because of such factors as production capacity and the lead time required to identify viral strains there will not be sufficient amounts of antiviral drugs or vaccines to protect all those potentially at risk during an influenza pandemic (although recent increases in production capacity may limit shortages of one key antiviral drug). The existence of such shortages will require that the distribution of these limited resources be prioritized. Traditionally, interventions have been distributed to those individuals most at risk on the principle of attempting to limit serious illness and death as much as possible. During a pandemic, however, preserving the functioning of society may be a higher priority, which would require that those individuals who are essential to the provision of health care, the maintenance of public safety, or the functioning of key aspects of society receive priority in the distribution of vaccines, antivirals, and other scarce resources. Such an approach will require that diverse stakeholders be involved in affirming this priority, determining who is deemed essential, and establishing a distribution strategy.

There are a variety of other ethical considerations identified by the Ethics Subcommittee. There should, for instance, be a commitment to transparency throughout the influenza planning and response process. Furthermore, since public engagement and involvement are essential to building public will and trust, they should be evident throughout this process. Public health officials have a responsibility to maximize preparedness in order to minimize the necessity for making allocation decisions later, during the course of the pandemic. Ethical guidelines should be based on the best available scientific evidence, with the current knowledge base serving as a foundation for these guidelines. There should

[16]The ethical guidance described in this article is based on the document developed by the Ethics Subcommittee of the Advisory Committee to the Director, CDC, *Ethical Guidelines in Pandemic Influenza*. The ethical guidelines document was prepared by Robert J. Levine and Kathy Kinlaw with input from other members of the Ethics Subcommittee and with assistance form Drue Barrett. The information was presented by James W. LeDuc to the IOM on September 20, 2006. A copy of *Ethical Guidelines in Pandemic Influenza* may be obtained at http://www.cdc.gov/od/science/phec/.

TABLE 3-1 Ethical Guidelines in Pandemic Influenza

General Ethical Considerations
Identification of clear planning goals
Commitment to transparency
Public engagement and involvement
Maximizing of preparedness
Sound guidelines based on best available scientific evidence
Global involvement and cooperation
Balancing of individual liberty and community interest
Diversity in ethical decision making
Fair process (procedural justice)

SOURCE: CDC (2007).

also be a commitment to ongoing scientific and ethical evaluation of interventions. The pandemic planning process should acknowledge the importance of working with and learning from preparedness efforts globally. The reasons for this collaboration include not only the potential of global involvement to benefit U.S. citizens (an "instrumental" reason) but also a recognition of global interdependence and the value of the common good.

It will be important in planning for pandemic influenza to balance individual liberties with community interests. Limits on individual freedom may be necessary to protect the community as a whole as well as those individuals whose liberty is restricted, yet individual liberty should be restricted with great care and only when alternative approaches are unlikely to be effective. In determining these restrictions, the guiding principles should include adopting the least restrictive practices, ensuring that restrictions are necessary and proportional to the need for protection, and ensuring that those affected by restrictions receive support from the community, such as job security and provision of necessities for the individuals and their families. Diverse public voices should be involved in determining the need for restrictions and in articulating their ethical justifications. Furthermore, planning and implementation should be done by decision makers who are impartial and neutral and who are consistent in applying standards, and those affected by the decisions should have a voice in making them—and, where feasible, agree in advance to the process. All who are affected by these decisions should be treated with dignity and respect.

Another important component of pandemic influenza planning will be resource allocation. The guidelines suggest that resource allocation should be designed to accomplish clearly articulated goals and be guided by criteria specified well in advance of a pandemic. The classic utilitarian approach of the greatest good for the greatest number is not appropriate for defining priorities in a pandemic influenza. Instead, the recommended approach is one that resembles utilitarianism in that it evaluates policies primarily in terms of their anticipated consequences but is tempered by the ethical principles of respect for persons, nonmaleficence, and justice. Distribution plans should specify what scarce goods are involved, who is to decide about prioritization and distribution, who is eligible to be a recipient, and what moral criteria will be used to assign priorities to groups of individuals. Criteria that would generally *not* be ethically supported include "to each according to purchasing power," "first come, first served," or criteria such as race, ethnicity, religious belief, or any similar characteristics used to make discriminations that are invidious and not morally relevant. Normally, distribution based on an individual's social worth is not morally acceptable; however, in planning for a pandemic, where the primary objective is to preserve the function of society, it may be necessary to identify certain individuals and groups as key to the preservation of society and to accord them a higher priority.

Social distancing and restrictions on personal freedom will be important tools for managing pandemic influenza. Such interventions can include the isolation of infected individuals; the quarantine of those heavily exposed, such as family members or close contacts; adjustments to school schedules or even the closing of schools and cancellation of public events; limiting travel; and restricting access to public venues. These interventions are founded on the premises that an individual sick with influenza is infectious only for a short time and that separating that person from the larger group of susceptible individuals during that time will likely interrupt further transmission of the disease. Putting any of these interventions into effect will involve restricting highly valued personal freedoms, so justification for any such restrictions must be carefully considered. The process for making decisions about these restrictions should be well thought out in advance and be done in a transparent manner by a group that is representative and diverse. Recent modeling of the effects of social-distancing measures at different stages of a pandemic suggests that voluntary compliance may be enough for such measures to succeed (IOM, 2006). Mandatory liberty-limiting and social-distancing interventions should be imposed only in situations where voluntary actions seem unlikely to be effective.

During a pandemic, centralized decision making will be necessary in a number of areas. Because this type of decision making represents a departure from customary public health practice, it will be important to create fair and equitable restrictions, and a process should be in place for objections to be heard, restrictions appealed, and new procedures to be considered. Local autonomy in decision making should be honored whenever there is no evidence that centralized decision making will contribute substantially to preserving the functioning of society and where the easing of restrictions is proportional and reasonable in particular communities (e.g., uniform duration of school closings may not be reasonable in communities where the influenza wave has already ended.) Communications about restrictions should begin early in the planning process, and the public should be clearly informed that restrictions on personal freedom are expected. Any liberty-limiting measure should be enacted only if the best available scientific evidence indicates that implementing the measure will achieve its intended goal, that the limitation is proportional to the anticipated benefit, that no less restrictive measure is likely to be effective, and that failure to implement the measure is likely to result in grave harm to the functioning of society. Throughout the process, the need for limits on individual freedom must continue to be assessed and affirmed.

Legal Considerations

Many legal challenges will undoubtedly emerge during the course of the global response to a pandemic of influenza. These will vary among countries and, domestically, across states, provinces, and localities in response to variation in the laws, to officials' competency in applying them, and to citizens' willingness to comply with, or challenge, the legal mandates (Fidler and Cetron, 2007; Stier and Goodman, in press). The CDC Public Health Law Program has described the concept of public health legal preparedness for public health emergencies, such as pandemic influenza, and it has identified four core elements that affect public health legal preparedness: a) laws; b) competencies in applying these laws; c) coordination of legal powers across jurisdictions and sectors; and d) information about best practices in implementing law-based interventions (Moulton et al., 2003; Goodman et al., 2006). These elements are highly relevant to social distancing, for instance, and steps that might be taken to maximize the effectiveness of social-distancing measures are described in detail elsewhere in this report (Cetron and Landwirth, 2005, and reprinted earlier in this chapter).

Experts generally agree that, as part of their police powers, all states have legal authority to quarantine and isolate individuals. The specific authorities and abilities of given states and other jurisdictions to quarantine groups, suspend public meetings, close facilities, and impose curfews are less well known. Experts differ on whether adequate social distancing can be achieved voluntarily. Table 3-2 lists a variety of issues related to public health law that will influence the effectiveness of mandatory social distancing, arranged according to the framework of the four public health legal-preparedness core elements.

Table 3-3 presents a number of questions about and challenges relating to the level of legal preparedness for social distancing. For example, what are the status and adequacy of laws for isolating those known to be, or strongly suspected of being, infected with pandemic influenza? Are public health officials legally authorized to quarantine those who have been in close contact with infected individuals? What is the status of legal authority to close schools and public gatherings? Can commercial movements and travel be restricted legally? To what extent? Do laws confer immunity to liability for health-care providers who, during a declared public health emergency, perform services for which they are not licensed or against whom claims of negligence may be asserted? Are the pertinent school laws uniform across adjoining communities? Do states have legal authority to assist in enforcement of a federal quarantine?

TABLE 3-2 Selected Public Health Law-Related Issues and Needs for Effective Mandatory Social Distancing

Legal Authorities to:
- Quarantine/isolate individuals and groups
- Modify the schedules of or close schools and public meetings
- Restrict commercial movement

Public and Private Officials Competent in:
- Application of social-distancing legal powers
- Protection of individual and property rights
- Legal responsibilities of health-care providers

Coordination of Legal Tools Across Jurisdictions and Sectors:
- Public health coordination with emergency response and law-enforcement agencies
- Public health coordination with health-care providers
- Coordination of social-distancing measures across communities and states

Information Resources on:
- Legal best practices in social distancing
- Effective communication of legal basis for social distancing with the public and the media

TABLE 3-3 Selected Challenges to Legal Preparedness for Social Distancing

Legal Authorities:
- Landscape of state/local laws is incompletely known
- Unclear if laws (e.g., school closure) are uniform
- Concern that laws may not provide due process, civil liberties, and property-rights protections

Competencies:
- Ensuring that officials of public health and other agencies are trained in use of social-distancing legal powers
- Ensuring that private health-care providers understand their legal responsibilities during pandemics
- Ensuring that public and private officials participate in tests of social-distancing legal preparedness

Coordination:
- Unclear if states can assist enforcement of federal quarantine
- Lack of protocols—e.g., between public health, law enforcement, and health care—for coordinated, cross-sector response

Information Resources:
- Guidance on sectors' roles and responsibilities
- Guidance for communicating with the public and the media

With respect to legal-preparedness competencies: Are officials trained in the application of legal powers to put social-distancing interventions into effect? Are they adequately prepared to protect individual and property rights while implementing social-distancing interventions? Are judges aware of the legal powers that public health officials hold and of the legal precedents relevant to appeals that aggrieved citizens may put forward? Do hospital executives and other health-care providers (and their legal counsel) understand their legal responsibilities during an influenza pandemic? Are officials capable of effective communication with the public and the media about the need for, and legal basis of, social-distancing interventions?

As to the coordination of legal powers across jurisdictions and sectors, there are concerns that the laws of some states may not allow those states to assist in enforcing a federally declared quarantine of, for example, passengers arriving on an international flight. It is thought that few jurisdictions have protocols in place for a coordinated response by health-care, law enforcement, and public health agencies to an infectious disease outbreak. A notable exception is the tripartite agreement executed in 2004 by the New York City health and police departments and the Federal Bureau of Investigation (FBI) for joint investigations of suspected bioterrorist attacks.[17]

Regarding the core element of information on legal best practices, there are a number of important issues: Do public health, health-care, law enforcement, and other relevant officials have ready access to such information in deciding when to use legal authorities to support social distancing? Are judges informed about the specific legal powers that public health officials possess? Do the executives and legal counsels of private hospitals know about these powers and understand their implications for hospital operation during emergencies?

These core elements of public health legal preparedness for pandemic influenza should be tested in every community and state by conducting exercises and other approximations of an actual pandemic. Such tests can help local and state officials and their private-sector counterparts identify gaps in legal authorities for mandatory social distancing, in case it should be needed, and also help them ascertain whether protocols are in place to translate those powers into practice. Exercises can test whether information is available to all the relevant government and private organizations concerning their legally specified roles and responsibilities during a pandemic, including their communications with the public and the media. Exercises also should test how well the applications of nonpharmaceutical interventions are coordinated between the federal government and the states, across states, and throughout jurisdictions within each state.

These and other legal aspects of the preparation for and response to pandemic influenza offer options for the Institute of Medicine and other organizations to participate further. Such participation might include, for example, efforts to assist

[17]Document available at http://www2.cdc.gov/phlp/docs/Investigations.PDF.

local, state and federal agencies, including CDC and the U.S. Department of Justice, in identifying gaps in the four public health legal-preparedness core elements in states and communities and then working to strengthen those core elements. In this effort, federal agencies should partner with state and local public health leaders and their legal counsel, their health-care counterparts, the courts, and all others who will play critical roles in shaping, implementing, and adjudicating the response to pandemic influenza.

Closing Thoughts

The threat of pandemic influenza has focused national and global attention on public health and on the tools that will be required to address this serious global challenge, some of which have not been used extensively in modern times. Because the scientific data needed for informed decision making are incomplete, models have been used extensively to predict outcomes based on representative scenarios of an influenza pandemic. These increasingly sophisticated models have proven valuable in exploring the possible outcomes of various policy decisions (IOM, 2006). One aspect of model development is careful documentation of assumptions made while building the model. These assumptions allow model builders to estimate various possible outcomes in the absence of hard data. It is essential that policy makers pay close attention to the assumptions underlying the models being developed and to the basis for these assumptions. Well-founded assumptions can guide future research aimed at pinpointing the key elements in our intervention strategies, and careful refinement of the assumptions will yield models that more faithfully represent reality.

Although a great deal of money and effort has gone into purchasing antiviral drugs for the treatment of pandemic influenza, and although good progress is being made in vaccine development and production, there remains a strong likelihood that these and other important items will be in short supply, and this will necessitate hard choices as to who receives them and who does not. Thus there needs to be extensive discussion, planning, and preparation concerning the allocation decisions and liberty-limiting and social-distancing interventions that will certainly be necessary in a pandemic. These steps should be taken only if the best available scientific evidence implies that they are likely to be successful, if they are grounded in sound ethical and legal principles, and if the important decisions about them are taken through a process that is transparent, inclusive, and appropriately communicated to the general public.

PUBLIC HEALTH AND ETHICAL CONSIDERATIONS IN PLANNING FOR QUARANTINE[18]

Martin Cetron, M.D.[19]
Centers for Disease Control and Prevention[20]

Julius Landwirth, M.D., J.D.[21]
Yale University

Quarantine is one of the oldest, most effective, most feared, and most misunderstood methods of controlling communicable disease outbreaks. Its etymological roots are traceable to fourteenth century public health practices requiring ships arriving in Venice from plague-infected ports to sit at anchor for 40 days (hence, quar-antine) before disembarking their surviving passengers. While in recent times the use of quarantine has been more humane and scientifically based, the historical association with exile and death and the morally negative connotation of sacrifice of a few for the benefit of others remains as an undercurrent of public apprehension. Nevertheless, quarantine was recently implemented successfully in several countries as a socially acceptable measure during the SARS epidemic in 2003 (Cetron and Simone, 2004). It is an important component of the Department of Health and Human Services (HHS) Pandemic Influenza Plan issued in November, 2005 (HHS, 2006a).[22]

The purpose of this article is to review the modern public health approach to quarantine, outline highlights of current plans for its implementation in the event of an avian influenza pandemic, and consider the ethical principles that should be considered.

Definitions

Quarantine is the restriction of persons who are presumed to have been exposed to a contagious disease but are not ill. It may be applied at the individual, group, or community level and usually involves restriction to the home or designated facility. Quarantine may be voluntary or mandatory.

Isolation is the separation of ill persons with contagious diseases. It may be applied at the individual, group, or community level.

[18]Reprinted with permission from *Yale Journal of Biology and Medicine* (2006). Copyright 2006 by *Yale Journal of Biology and Medicine*.

[19]Division of Global Migration and Quarantine.

[20]The findings and conclusions in this manuscript have not been formally disseminated by CDC and should not be construed to represent any agency determination or policy.

[21]Interdisciplinary Center for Bioethics and Donaghue Initiative in Biomedical and Behavioral Research Ethics.

[22]This plan was issued after the Yale symposium on Ethical Aspects of Avian Influenza Pandemic Preparedness but is included in these published proceedings for completeness.

Quarantine of groups refers to quarantine of people who have been exposed to the same source of illness (e.g., at public gatherings, airline, school, workplace).

Working quarantine refers to persons who are at occupational risk of influenza infection, such as health-care workers, who may be restricted to their homes or designated facilities during off duty hours.

Community-wide quarantine refers to closing of community borders or the erection of a real or virtual barrier around a geographic area (cordon sanitaire).

Modern public health places quarantine within a broader spectrum of interventions generally referred to as "social distancing."

The effect of successful measures to increase social distance is to convert a dynamic of exponentiation in the spread of an infectious agent to one of suppression in which the number of secondary cases from exposed persons is reduced to a manageable level. Time is the key variable in the success or failure of social distancing strategies, including the duration of communicability, whether or not communicability occurs before onset of symptoms, the number of resulting contacts, and the efficiency of or delays in contact tracing.

Globalization of travel and trade, and decreased travel time between distant places have further complicated these relationships. There are several hundred international ports of entry airports in the United States. Fortunately, 25 of these airports account for approximately 85 percent of international arrivals. Detailed recommendations for travel-related containment measures can be found in the full HHS report and will not be further elaborated here.

Principles of Modern Quarantine

In the months before adequate supplies of vaccines and antiviral agents are expected to be available, quarantine and isolation are likely to be the mainstays of containment strategies.

> The HHS plan states that: The goal of quarantine is to protect the public by separating those exposed to dangerous communicable disease from the general population. It represents collective action for the common good that is predicated on aiding individuals who are already infected or exposed and protecting others from inadvertent exposure (HHS, 2006b).

Principles of modern quarantine and social distancing limit their use to situations involving highly dangerous and contagious diseases and when resources are reliably available to implement and maintain the measures. It encompasses a wide range of strategies to reduce transmission that may be implemented along a continuum based on phase and intensity of an outbreak.

For example, at a stage when transmission of a novel influenza virus is still limited, either abroad or in the area, and local cases are either imported or have clear epidemiological links to other cases, individual quarantine of close contacts may be effective. At a more advanced phase of the pandemic, however, when virus transmission in the area is sustained and epidemiological links to other known cases is unclear, limiting quarantine to exposed individuals may be ineffective, and the strategy may need to expand to include community-based interventions that increase social distance. These include school closings, cancellation of public gatherings, encouraging non-essential workers to stay home, and reduced holiday transportation schedules. If these measures are believed to be ineffective, community-wide quarantine may need to be implemented.

The HHS guidelines cite two important principles designed to help ensure that those in quarantine are not placed at increased risk. First, quarantined individuals will be closely monitored, with daily visits as needed, in order to detect earliest onset of symptoms and separation from those who are well. Second, persons in isolation will be among the first to receive any disease-prevention interventions. In addition, the HHS plan recommends that they should be provided with all needed support services, including psychological support, food and water, and household and medical supplies.

Home quarantine is the preferred method of separation, whenever possible. Designated quarantine facilities may have to be identified for potentially affected persons who do not have access to an appropriate home environment, such as persons living in dormitories, travelers, the homeless, or if the configuration of the home is not suitable for the protection of the potentially infected person and other occupants.

Voluntary quarantine is the preferred first option before resorting to mandatory orders or surveillance devices. In this connection, it is noteworthy that quarantine does not require 100 percent compliance to be effective. Toronto Public Health officials reported only 22 orders for mandatory detainment among the approximately 30,000 persons who were quarantined (Upshur, 2003).

Legal and Ethical Considerations

Primary responsibility for public health matters within their borders rests with state and local governments. This includes isolation and quarantine. Applicable state laws, regulations and procedures vary widely. A recently developed Model State Emergency Health Powers Act attempts to promote greater interstate consistency in response to emergency public health situations (Center for Law and the Public's Health, 2001). In the section on isolation and quarantine, the Model Act covers the principles and conditions governing implementation of quarantine; authorization of public health authorities to impose temporary quarantine by directive, with rights of appeal within 10 days; imposition of quarantine with notice following a public health authority court petition and hearing; and legal procedures for release from quarantine or relief from violations of conditions of quarantine. Although it has been criticized by some as being overly broad in its coercive powers (Annas, 2002; Mariner, 2001) the Model Act has been adopted in whole or part in a number of jurisdictions.

The federal Public Health Service Act (U.S. Congress, 1946) gives the HHS secretary responsibility for preventing introduction, transmission, and spread of communicable diseases from foreign countries into the United States and within the United States and its territories/possessions. This authority is delegated to the Centers for Disease Control and Prevention (CDC), which are empowered to detain, medically examine, or conditionally release individuals reasonably believed to be carrying a communicable disease. The Public Health Service Act also provides that the list of diseases for which quarantine is authorized must first be specified in an executive order of the president, on recommendation of the HHS secretary (CDC, 2006). On April 5, 2005, influenza caused by a novel or reemergent strain that is causing or has the potential to cause a pandemic was added to that list (White House, 2005).

Although the discipline of public health has its origins several centuries ago, it is only relatively recently that ethical principles and codes to guide public health practice and policy have been formulated. The ethical principles at the heart of the more fully developed fields of medical and research ethics are grounded in the primacy of individual autonomy in clinical decision-making in the therapeutic setting and in consent for participation in the setting of human subjects research. They are guided by a fundamental moral axiom that individual persons are valued as ends in themselves and should never be used merely as means to another's ends. Public health, on the other hand, emphasizes collective action for the good of the community.

The Principles of the Ethical Practice of Public Health, issued by the Public Health Leadership Society in 2002 (Public Health Leadership Society, 2002), states that community health should be achieved in a way that respects the rights of individuals and the community. Accompanying notes are instructive:

> This principle identifies the common need in public health to weigh the concerns of both the individual and the community. There is no ethical principle that can provide a solution to this perennial tension in public health. We can highlight, however, that the interest of the community is part of the equation, and for public health it is the starting place in the equation; it is the primary interest of public health. Still, there remains the need to pay attention to the rights of individuals when exercising the police powers of public health (Public Health Leadership Society, 2002).

To address this potential dichotomy, the principles require ensuring opportunity for informed community participation in the development of policies, programs, and priorities, accessibility to basic resources and conditions necessary for health, and protection of confidentiality.

Principles of practice, law and ethics in the containment of outbreaks of infectious disease, especially use of quarantine, confront a common underlying concern, namely,

> The individual fear and community panic associated with infectious diseases often leads to rapid, emotionally driven decision making about public health policies needed to protect the community that may be in conflict with current bioethical principles regarding care of individual patients (Smith et al., 2004)

In November 2005, the Council on Ethical and Judicial Affairs of the American Medical Association issued recommendations for the medical profession in the use of quarantine and isolation as public health interventions. Again, the tensions between the ethical imperatives of therapeutic medicine and public health are reflected in the following excerpts:

> Quarantine and isolation to protect the population's health potentially conflict with the individual rights of liberty and self-determination. The medical profession, in collaboration with public health colleagues, must take an active role in ensuring that those interventions are based on science and are applied according to certain ethical considerations. . . . Individual physicians should participate in the implementation of appropriate quarantine and isolation measures as part of their obligation to provide medical care during epidemics. . . . In doing so, advocacy for their individual patients' interests remain paramount (Council on Ethical and Judicial Affairs, 2005).

An important rationale for acknowledging and attempting to ameliorate this tension in pandemic preparedness planning, including quarantine measures, is to reduce the potential for unfair distribution of burdens and benefits among various segments of society (Markovits, 2005). In an important contribution, Susan Kass has developed a six-step framework for ethical analysis specifically for public health (Kass, 2001). The application of this general framework to quarantine is discussed in detail elsewhere in these proceedings.

Ross Upshur has outlined four principles that must be met to justify quarantine (Upshur, 2002):

> First, under the harm principle there must be clear scientific evidence of person-to-person spread of the disease and the necessity of quarantine as a containment measure. Second, the least restrictive means should be implemented. Third, upholding the principle of reciprocity points to the community's obligation to provide necessary support services for those in quarantine. Fourth, the obligation of public health authorities is to communicate the reasons for their actions and to allow for a process of appeal. In November 2004, the World Health Organization issued a checklist for influenza pandemic preparedness. It encourages planners to "consider the ethical issues related to limiting personal freedom, such as may occur with isolation and quarantine" (WHO, 2005a).

An instructive example of how ethical considerations can be incorporated into pandemic preparedness plans can be found in the Ontario Health Plan for an Influenza Pandemic (Ontario Ministry of Health and Longterm Care, 2005). The development of this plan included a collaboration with the Toronto Joint Centre for Bioethics, which produced a 15-point ethical guide for decision making for a pandemic (University of Toronto Joint Centre for Bioethics Pandemic Influenza Working Group, 2005). The guide identified four key ethical issues in pandemic preparedness planning, one of which was "restricting liberty in the interest of public health by measures such as quarantine." The guide describes the following substantive and procedural ethical values at stake in addressing this issue:

1. Individual liberty: Isolation and quarantine should be proportional, necessary, relevant, equitably applied, and done by least restrictive means.
2. Protection of public from harm: Officials must weigh the imperative for compliance and review decisions.
3. Proportionality: Restrictive interventions should be limited to the actual level of risk to community.
4. Privacy: There must be a necessity for overriding the public's protection.
5. Reciprocity: Support is needed for those facing a disproportionate burden in protecting public health, individual liberty (proportional, necessary, relevant, least restrictive means, equitably applied), protection of public from harm (weigh the imperative for compliance, review decisions), proportionality (restrictive interventions limited to actual level of risk to community), privacy (necessity for overriding for public's protection), and reciprocity (support for those facing disproportionate burden in protecting public health).

Procedures should be reasonable, with reasons for decisions shared with stakeholders; open and transparent; inclusive, with stakeholder participation; responsive, subject to review and revision with experience; and accountable.

Based on these principles, the guide recommended the following:

1. Governments and the health-care sector should ensure that pandemic influenza response plans include a comprehensive and transparent protocol for the implementation of restrictive measures. The protocol should be founded upon the principles of proportionality and least restrictive means, should balance individual liberties with protection of public from harm, and should build safeguards such as the right of appeal.
2. Governments and the health-care sector should ensure that the public is aware of a) the rationale for restrictive measures, b) the benefits of compliance, and c) the consequences of noncompliance.
3. Governments and the health sector should include measures in their pandemic influenza preparedness plans to protect against stigmatization and to safe-

guard the privacy of individuals and/or communities affected by quarantine or other restrictive measures.

4. Governments and the health-care sector should institute measures and processes to guarantee provisions and support services to individuals and/or communities affected by restrictive measures, such as quarantine orders during a pandemic influenza emergency. Plans should state in advance what backup support will be available to help those who are quarantined (e.g., who will do their shopping, pay the bills, and provide financial support in lieu of lost income). Governments should have public discussions of appropriate levels of compensation in advance, including who is responsible for compensation.

Past experience has shown that voluntary cooperation and public trust are key ingredients of successful response to a public health emergency. They may be important antidotes to individual fear and community panic that may be engendered by infectious disease outbreaks. Careful attention to the ethical values at stake in public health decision making can help foster voluntary cooperation and public trust and should be a part of state and federal pandemic preparedness planning.

REMARKS ON THE ROLE OF MODELING IN INFECTIOUS DISEASE MITIGATION AND CONTAINMENT

Joshua M. Epstein, Ph.D.[23,24]
The Brookings Institution

I have been invited to make some brief informal remarks about the modeling enterprise and its relevance to infectious disease mitigation and containment. I will try to be brief, and will certainly be informal, in the sense of dispensing with extensive citations. Following this paper are two published articles which serve as recent examples of my own modeling. I would refer you to these articles for complete references and technical details.

Implicit and Explicit Models

I would like to begin with a claim that most audiences find rather discomfiting. It is this: *Everyone in the room is a modeler.* You are all modelers. Anyone

[23]Senior Fellow, Economic Studies Program and Director, Center on Social and Economic Dynamics, The Brookings Institution. Also affiliated with the National Institutes of Health (NIH) Models of Infectious Disease Agent Study (MIDAS). MIDAS is a large multi-institution modeling project funded by the National Institute of General Medical Sciences (NIGMS). This chapter is an edited version of the author's address to the National Academy of Sciences, Institute of Medicine's Forum on Microbial Threats, September 20, 2006.

[24]The author thanks Ross Hammond for insightful comments on these remarks.

who ventures a projection, or imagines how things might unfold in space and time, is running *some* model or other. It is just an *implicit* model in which the assumptions aren't actually laid out, in which their internal consistency can't really be tested, in which their consequences can't be played out and examined very rigorously, in which their relation to data is unknown. But when you close your eyes and imagine an epidemic spreading, or any other social phenomenon, you are running some model or other. It is just an implicit model that you haven't written down.

I am always amused when smug practitioners challenge me with the question: "Can you validate your model?" The appropriate retort, of course, is: "Can you 'validate' yours?" At least I can write mine down so that it can, in principle, be calibrated to data (if that is what you mean by "validate").

Accordingly, I build *explicit* models, and we at NIH/MIDAS build explicit models, so that we can study exactly what our assumptions entail: On these assumptions, *this* sort of thing happens. When you alter the assumptions, *that* is what happens.

When you write *explicit* models you let others replicate your results. You *can* calibrate the model to historical cases, if there are data. You can test against current data to the extent that exists. And, importantly, you can incorporate the best medical expertise in a rigorous way.

With modern computing, the models, if need be, can be quite realistic spatially, behaviorally, biomedically, and in numerous other ways. You can execute a huge range of possible scenarios and containment strategies and do what we call *sensitivity analysis* to identify the most salient uncertainties. You know there are going to be uncertainties and you want to know how important they are. Which ones can you gauge? About which ones do you need to learn more?

Model Myths

There are some myths about explicit models that I want to refute immediately. Certainly in the policy sphere (if not in particle physics), such models do not replace judgments. They can make judgments better informed. They permit us to incorporate the best expertise and existing data in a rigorous way. But in making public health policy, there will be a central role for judgments, and models[25] (at least at this stage of development) don't change that.

Models, likewise, do not eliminate uncertainty. We are going to face uncertainty. If we could bound it, that would be useful. Sometimes models can help us do that, and help us identify those uncertainties that actually matter most. Not all uncertainties are equal. Models can help us rank them and set priorities and even figure out what data we need to collect. Without a model it is often not clear exactly what data are worth collecting.

---

[25]I will not always repeat the term "explicit," as it will be clear from the context.

On this point, non-modelers (and many modelers) often adopt a naive inductivism: "Science proceeds from observation, and then models are constructed to 'account for' the data." Not always so. Maxwell's theory predicted the existence of radio waves, which were only later sought, and observed. Further examples abound.[26]

While purely theoretical work is essential to basic scientific progress, in *applied* infectious disease modeling I think it is very important that modelers work closely with medical experts. This is one of the strengths of the MIDAS operation: when we set about trying to model pandemic flu, we actually had intensive consultations at CDC with the best flu experts we could convene, to thrash out what we should actually assume about a human-to-human variant of H5N1. Same for smallpox. We worked very closely with D.A. Henderson and others to arrive at detailed biomedical and behavioral assumptions for normal, modified (by prior immunity), and hemorrhagic smallpox before comparing intervention strategies (see below).

Just delivering a mathematical black box with lots of dials to twiddle—as we modelers are prone to do—is really not useful to decision makers. You have got to roll up your sleeves together with actual domain experts and grapple with everything, sort it out, and make your assumptions clear. It's very useful that computers are such dumb, literal beasts. If you don't make your assumptions crystal clear to the computer—and therefore to yourself—it just won't run. The modeling enterprise itself forces this kind of explicitness upon you, and it may be as valuable as the final model.

From Ignorant Militance to Militant Ignorance

In particular, the enterprise of modeling—of squarely facing one's imprecisions and uncertainties—engenders a healthy humility. At its core, the scientific habit of mind involves a kind of *militant ignorance*[27]; that's right, an insistence on "I don't know." Much more common, I'm afraid, is an *ignorant militance;* a commitment to one's preconceptions despite evidence.

A Rose Is Not a Rose

So, having convinced you, maybe, that modeling per se isn't so bad, I want to say that there are profoundly different types of models. One type of model that has dominated the landscape until quite recently is the traditional so-called "low dimensional ordinary differential equation model." These are very widely used

---

[26]Einstein's Theory of General Relativity predicted the deflection of light by gravity, which was only later confirmed by experiment.

[27]On this point, see Richard P. Feynman, 1998. *The Meaning of It All: Thoughts of a Citizen-Scientist.* Reading, MA: Perseus Books. Pp. 26-28.

to project epidemic severity, to compare intervention strategies, and to estimate demand for vaccine, medical personnel, and so forth.

The problem with these models is that they typically assume a kind of perfect mixing in the population, what in physics would be called *mass action kinetics*. It is as though you took infectives and susceptibles and put them in a jar and shook the jar up with tremendous energy so that everybody smashes into everybody as if they were particles in a gas. In turn, the particles, if you will, are themselves homogeneous. Typically, it is just susceptibles and infectives. There is very little differentiation *among* the susceptibles or among the infectives in these models.

All of this makes for very elegant mathematics and has produced absolutely fundamental insights (e.g., herd immunity) and is entirely defensible in some domains. But in twenty-first-century urban industrial settings it can very seriously distort our estimates of severity and the policies we base on those estimates.

The real world is just a lot more complicated. It is very far from perfectly mixed. How do things unfold if smallpox or pandemic flu or some other bug is released on a modern metro system or at an international airport? In such cases, well-mixed models can be seriously misleading. Models are idealizations, and I am all for that, but they have to be productive and not misleading idealizations; to assume perfect mixing in processes of the sort we are worrying about here is really not productive. Indeed, I think it can do some damage in fact.

At the Brookings Center on Social and Economic Dynamics and in MIDAS, my colleagues and I are building models of a fundamentally different sort. We call them individual-based, or *agent-based*, models (ABM), and the core idea is this: Rather than write aggregate equations for the uniform mixing of entire homogeneous pools, in ABMs every single individual is represented. These are completely disaggregated models. Every individual is represented as a discrete software object, a little "cyberperson" in his or her own right. These agents (not to be confused with disease agents) can differ from one another in myriad ways. They can differ by age, by immunocompetence, by disease state. They can have actual itineraries, agendas, and activities that they execute in the computer. They may commute to work, go to school, work at the hospital, travel someplace, and adapt to conditions over time. Events unfold not in a well-mixed jar but on an explicit landscape of some sort, a town, a country, the globe. As I will show you, there are actual social networks governing who bumps into whom.

The Smallpox Model

The smallpox model was initiated by me and Donald Burke,[28] then of the Johns Hopkins Bloomberg School, at the invitation of the National Academy of Sciences' Committee on International Security and Arms Control (CISAC). At the time, the CISAC's main concern was genetically modified variants, specifically

[28]Currently the Dean of the University of Pittsburgh Graduate School of Public Health.

IL-4 variants. This was before 9/11 and the anthrax attacks. Subsequently, of course, issues of smallpox bioterror rose to the fore, and we joined the Smallpox Modeling Working Group of the Secretary of Health and Human Services' Advisory Council on Public Health Preparedness. This working group was founded and chaired by D.A. Henderson, who also chaired the Advisory Council, to which we presented our final results. These are presented in detail in the articles attached with this lecture.

However, to reinforce my earlier point about the need for close collaboration between modelers and medical experts, the Working Group had intensive regular meetings to arrive at detailed assumptions about all the biomedical and critical behavioral aspects of this problem, and when I say "detailed assumptions," I really mean it.

Figure 3-1 shows the natural history assumptions we arrived at for ordinary smallpox. We also developed the natural history assumptions for smallpox modified by prior immunity, and for hemorrhagic smallpox. It required the better part of two years of work, of really rolling up our sleeves—with experts directly engaged in the smallpox eradication—to converge on these assumptions and write them down explicitly and carefully. The results are published now in the *International Journal of Infectious Diseases* (Longini et al., 2006, and page 112). In addition to accommodating highly detailed inputs, agent-based models produce quite novel outputs.

Dendrograms

For example, because they track each individual (unlike homogeneous differential equations), agent-based models permit us to reconstruct exact transmission chains and thereby to address the question (which loomed large for SARS) of what makes someone a superspreader. Is it the person's biology, his or her position in the social network, some foreseeable combination of factors? For the particular model realization depicted in Figure 3-2, hemorrhagics (red) die before they can transmit. But modified smallpox cases (green) —with less severe symptoms due to prior immunity—transmit very effectively, while the main transmitters are agents with ordinary smallpox (blue).

For a complete exposition of this model and, importantly, its calibration to historical data, please see the attached articles.

Let me emphasize again a practical advantage of agent-based models: Because they are rule-based rather than equation-based, they are accessible to non-modelers, which facilitates collaboration with medical experts. At the same time, there is no loss of rigor. The models can be run a large number of times under different assumptions and stochastic perturbations to produce a robust statistical portrait of the model's behavior, which can then be compared to data by appropriate statistical means.

Regarding our work with D.A. Henderson on smallpox, I think it is fair to

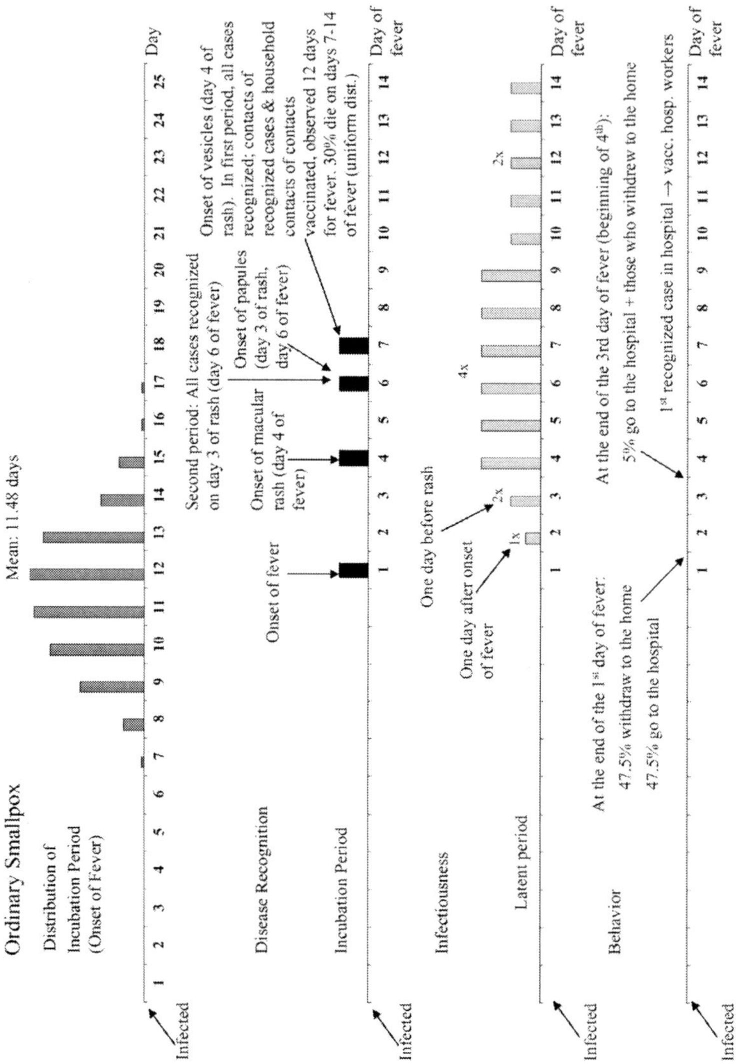

FIGURE 3-1 Ordinary smallpox natural history.
SOURCE: Longini et al. (2006).

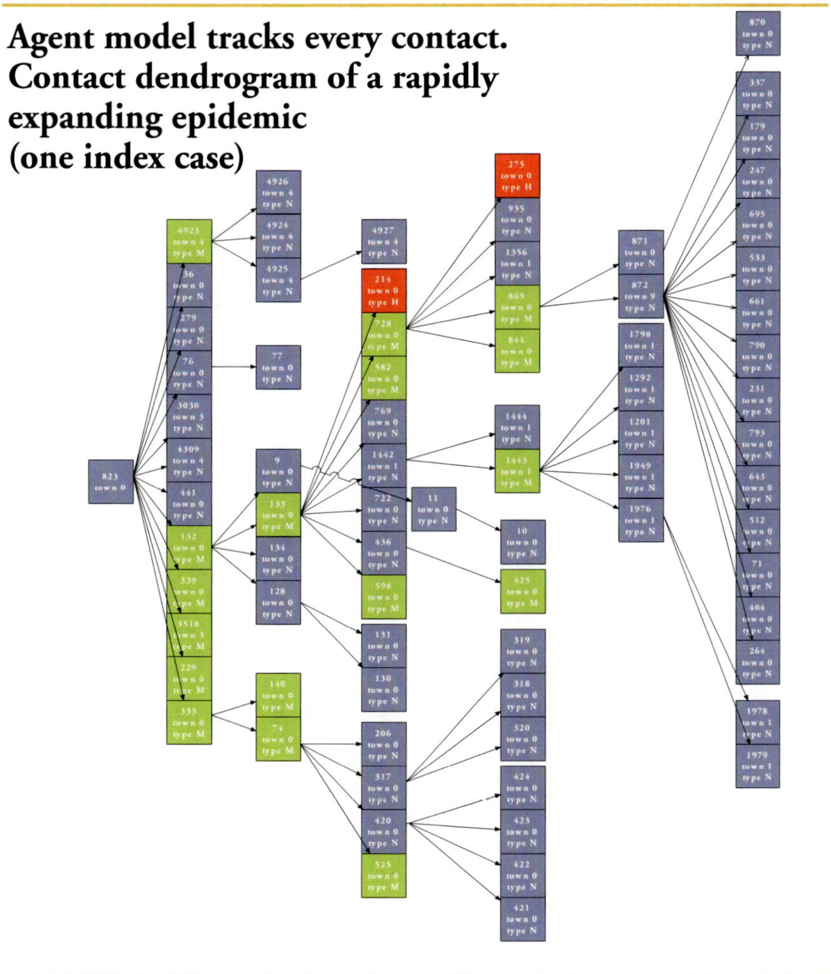

FIGURE 3-2 Contact dendrogram of a rapidly expanding epidemic (one index case).
SOURCE: Derived from Epstein et al. (2004).

say that, through engagement with domain experts, facilitated by a user-friendly agent-based approach, and incorporating all available data, we were able to develop novel and powerful approaches to containment; we met high scientific standards; and we enjoyed high policy credibility. Neither modeling in isolation nor expert opinion unsupported by models would have had the combined scientific and policy impact this collaboration achieved.

We aim to replicate this excellent experience on global pandemic flu within

MIDAS, and on a wide range of public health questions at the Brookings Center on Social and Economic Dynamics.

Concluding Points

To wrap up these brief informal remarks, I would say that explicit models are like Democracy. They are the worst imaginable system except for all the others, and in studying dynamics with the scale and complexity of global pandemic flu, I don't think there is any alternative to explicit models. There is also no alternative to judgment, and modelers really need to work closely with medical professionals to make models that are maximally useful. As I say, delivering a black box with knobs to twiddle may not be very helpful. My own experience is that interdisciplinary teams comprised of modelers and medical experts are really the way to go. That is the way I have been going, and I certainly look forward to going further on pandemic flu and other emerging public health challenges.

In conclusion, to quote the great statistician, George Box, "All models are wrong, but some are useful."

CONTAINING A LARGE BIOTERRORIST SMALLPOX ATTACK: A COMPUTER SIMULATION APPROACH[29]

Ira M. Longini, Jr., Ph.D.
Fred Hutchinson Cancer Research Center[30]
University of Washington[31]

M. Elizabeth Halloran, M.D., D.Sc., M.P.H.
Fred Hutchinson Cancer Research Center[30]
University of Washington[31]

Azhar Nizam, M.S.[32]
Emory University

Yang Yang[33]
Harvard University

Shufu Xu, M.S.[30]
Fred Hutchinson Cancer Research Center

---

[29]Reprinted with permission from Elsevier. Copyright Elsevier 2006. Original article was in *International Journal of Infectious Disease* 11(2):98-108.

[30]Program in Biostatistics and Biomathematics.

[31]Department of Biostatistics, School of Public Health and Community Medicine.

[32]Department of Biostatistics, Rollins School of Public Health.

[33]School of Public Health.

Donald S. Burke, M.D.[34]
Johns Hopkins University

Derek A. T. Cummings, Ph.D.[34]
Johns Hopkins University

Joshua M. Epstein, Ph.D.[35]
The Brookings Institution

Introduction

The intentional release of smallpox remains a threat to the American population (Cohen and Enserink, 2002; Henderson et al., 1999). Our earlier work (Halloran et al., 2002b) showed that for a small attack involving around five initial infectives, post-release targeted vaccination of close contacts of identified infected people would be sufficient to control the epidemic. Our result is supported by other investigators using a simpler model (Eichner, 2003a) There is modeling evidence that a large attack may be difficult to contain (Kaplan et al., 2002). A stochastic model of a large smallpox attack indicates that targeted vaccination combined with early detection may be effective without resorting to mass vaccination (Eubank et al., 2004). Other work, not based on a dynamic epidemic model, suggests that given the small probability of a bioterrorist smallpox attack, preemptive mass vaccination is not a good strategy as opposed to reactive containment strategies (Bozzette et al., 2003).

Our earlier work also showed that preemptive voluntary vaccination to increase herd immunity could increase the effectiveness of a surveillance and containment control strategy, but further investigation would be needed for the case of a large attack. Currently there is virtually no effort to vaccinate the civilian population in the USA. The federal government's goal to vaccinate 5–10 million first responders and hospital personnel preemptively by the summer of 2003 was not achieved (U.S. Census Bureau, 2000). Even the more modest plan to vaccinate 500,000 medical personnel has not been achieved (Enserink, 2003). As of October 31, 2005, about 40,000 people had been vaccinated (CDC, 2005c), and many states had paused their smallpox vaccination program pending further federal government guidance. Routine vaccination against smallpox in the USA was stopped in 1972, currently leaving at least 43 percent of the population of the USA completely susceptible. However, evidence suggests that substantial residual immunity remains in those previously vaccinated (Hammarlund et al.,

[34]Department of International Health, The Bloomberg School of Public Health. Currently, Dr. Burke is the Dean of the Graduate School of Public Health at the University of Pittsburgh.

[35]Center on Social and Economic Dynamics.

2003), and such immunity would give partial protection against severe disease and death given infection (Eichner, 2003b).

In this work, we address the question of whether post-release surveillance and containment, i.e., isolation of detected smallpox cases, and location and vaccination of their close contacts, would be sufficient to contain even a large smallpox release, given the current level of background immunity to smallpox in the population of the USA. We also examine the added benefit of prevaccination of hospital workers, reactive mass vaccination of the population after an attack has been detected, and reactive closing of the schools.

Materials and Methods

Many of the parameters and scenarios of our model were determined by the Smallpox Modeling Working Group, the Secretary's Advisory Council on Public Health Preparedness, Department of Health and Human Services.[36] Parameter values and modeling decisions made by the working group were based on the group's collective knowledge of smallpox epidemiology and on information from Chapter 4 of *Smallpox and its Eradication* by Fenner et al. (1988). The simulation model developed here is a direct extension of our previous model (Halloran et al., 2002b), but for a larger population and potential attack. In addition, through the working group, we were able to derive a more accurate set of natural history parameters than for the previous model. The smallpox natural history and human behavior patterns that we give in the next section represent a blending of values from the literature and expert opinion that may be the most comprehensive description up to this time.

Natural History, Behavior, and Control Methods

We described the natural history of smallpox in terms of three timelines (Figure 3-3): (1) disease symptoms and recognition, (2) infectiousness, and (3) behavior of infected people. We also partitioned smallpox cases into three categories: (1) ordinary smallpox (Figure 3-3), (2) modified smallpox (Figure 3-4), and (3) hemorrhagic smallpox (Figure 3-5). For those who have never been vaccinated, we assumed that 95 percent would develop ordinary smallpox if infected, and the remaining 5 percent would develop hemorrhagic smallpox if infected. For those people over 32 years of age who were vaccinated before 1971, we assumed that 10 percent would be fully protected against smallpox infection,

[36]The working group was headed by J Chin (University of California, Berkeley) and also consisted of L Anderson (CDC), L Borio (HHS), J Breman (Fogarty International Center, NIH), G Curlin (National Institute of Allergy and Infectious Diseases, NIH), J Donlon (HHS), E Eitzen (HHS), DS Burke (Johns Hopkins School of Public Health), JM Epstein (Brookings Institution), JW Glasser (CDC), ME Halloran (Emory University), DA Henderson (HHS), IM Longini (Emory University), E McKenzie (NIH/FIC), M Miller (NIH/FIC), F Murphy (University of California, Davis).

30 percent would develop a less severe modified case of smallpox if infected, and the remaining 60 would develop non-modified smallpox if infected. Among that 60 percent, 95 percent of the cases would be ordinary smallpox and 5 percent hemorrhagic smallpox. About 57 percent of the population of the USA was born before 1971. We divided our simulated outbreaks into two periods. The first period is before recognition of smallpox, while the second period is after the first case of smallpox is recognized.

Figure 3-3 shows the natural history for ordinary smallpox. The incubation period distribution was assumed to vary from 7 to 17 days with a mean of 11.48 days. The incubation period was assumed to end with the onset of fever, followed by a macular rash on the 4th day of fever, with subsequent onset of papules and then vesicles. Before smallpox is known to be present, smallpox cases would not be recognized as such until the onset of vesicles, seven days after the onset of fever. After smallpox is known to be present, cases would be recognized at the onset of papules, six days after the onset of fever. Thirty percent of the cases would die 7–14 days after the onset of fever. People have varying degrees of transmission capabilities over the course of their infectious period, as shown in Figure 3-3. According to this pattern, 92 percent of an infected person's infectiousness occurs after the onset of the macular rash, an assumption consistent with a recent statistical analysis of smallpox infectivity (Eichner and Dietz, 2003). Figure 3-3 shows that 47.5 percent of the cases would withdraw to the home at the end of the first day of fever, and 47.5 percent would go to the hospital at that time. The remaining 5 percent would continue to circulate but go to the hospital at the end of the third day of fever.

We modeled modified smallpox to have a similar incubation period to that of ordinary smallpox, but a milder course of disease with only a 10 percent case fatality rate (Figure 3-4). The infectiousness of people with modified smallpox would be 33 percent of that for people with ordinary smallpox. Hemorrhagic smallpox was modeled to have a shorter natural history and more severe disease progression than ordinary smallpox with a 100 percent case fatality rate (Figure 3-5). Infected people would begin internal bleeding four days after the onset of fever, and die on the seventh day after the onset of bleeding. Before smallpox is known to be present, we assumed that 50 percent of hemorrhagic smallpox cases would not be recognized and 50 percent would be recognized on the fifth day of fever. After smallpox is known to be present, all hemorrhagic cases would be recognized on the fourth day of fever. People with hemorrhagic smallpox would be five times more infectious than those with ordinary smallpox.

The Population

The model populations are based on a 50,000 person network of structured subpopulations of 2,000 people mixing in households, clusters of households, neighborhoods, preschool groups, schools, and the community at large. The age

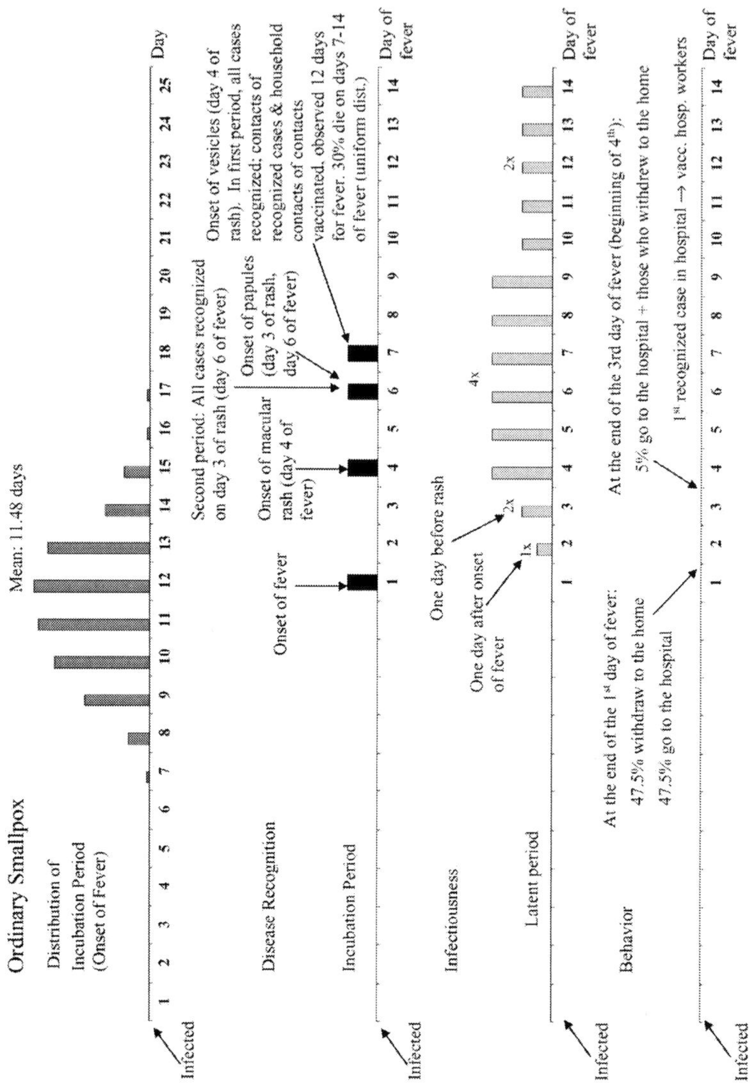

FIGURE 3-3 The natural history of ordinary smallpox in terms of time lines. Parameter values were determined through group consensus from the Smallpox Modeling Working Group and Chapter 4 of Fenner et al. (1988). The length of the incubation period follows the probability distribution shown in the top line. At the end of the incubation period, cases develop a fever, and then pass though a series of disease states. Before smallpox is recognized in the hospital (i.e., the first period), an ordinary case of smallpox would be recognized on the fourth day of rash. After this (i.e., the second period), smallpox is known to be present and all ordinary smallpox would be recognized in the hospital on the third

continued

distribution and approximate household sizes were based on the U.S. Census 2000 (U.S. Census Bureau, 2000). The subpopulations are connected through adult workplaces and high schools, and the whole population through a hospital. We include one hospital since statistics show there is about one hospital per 50,000 people in the USA.[37] The hospital has a total of 686 workers, 133 of whom can make close contact with smallpox cases until isolation measures in the hospital are instituted, based on a review of the numbers of employees having routine contact with patients.[37] Each person in the population may visit the hospital with probability 0.001 each day. They mix with all infected people in the hospital in the first period before smallpox is recognized, but only with unisolated circulating cases during the second period after smallpox is recognized. Figure 3-6A shows a schematic of the configuration of a subpopulation of 2,000 people. Figure 3-6B shows how the subpopulations are connected through schools, workplaces, and a hospital to form the population of 50,000 people.

The Simulation Model

We developed a discrete-time, stochastic simulation model of smallpox spread within a structured population described above. As mentioned above, the model is an extension of our previous smallpox model (Halloran et al., 2002a), but for a larger population and with a more detailed disease natural history description. The model represents the number of close and casual contacts that a typical person makes in the course of a day. The basic person-to-person daily transmission probabilities, x, and mixing group sizes are given in Table 3-4 (see Annex 3-1). We define x as the probability that an infected person with ordinary smallpox, on the second day after the onset of fever, makes sufficient contact to infect an unvaccinated susceptible person in the mixing group being modeled. For example, if a child were infected with ordinary smallpox, the probability that this child would infect an unvaccinated adult in the household, one day after the onset of fever, would be 0.05. On the third day after the onset of fever, this probability would increase to 0.10 (see infectiousness timeline in Figure 3-3). The transmis-

[37]Smallpox modeling working group, Secretary's Advisory Council on Public Health Preparedness.

---

day of rash. For infectiousness, the per contact transmission probability x (Table 3-4, see Annex 3-1), is set to 1x for the first day of fever, increased to 2x for the second day of fever, 4x at the onset of rash, etc., with an upper limit of 1.0. Thirty percent of ordinary smallpox cases would die between days 7 and 14, according to a uniform distribution. In the behavior time line, cases withdraw to the home or go to the hospital according to the pattern indicated. In surveillance and containment, close contacts of identified cases are vaccinated.

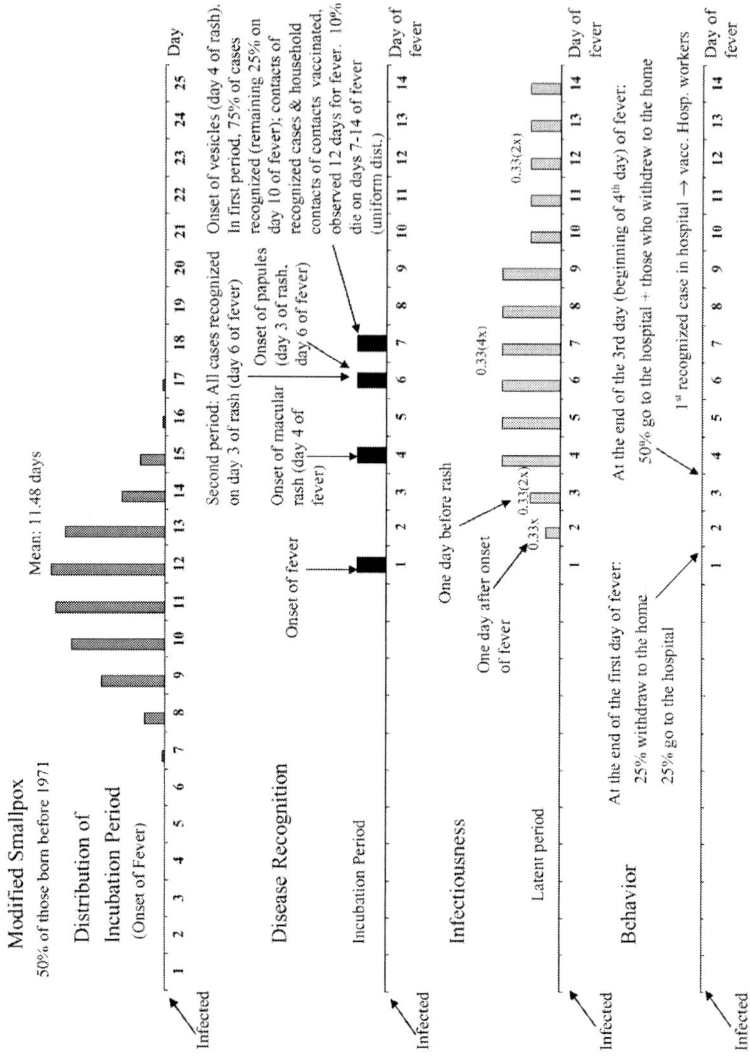

FIGURE 3-4 The natural history of modified smallpox in terms of time lines. Modified smallpox is assumed to have the same incubation period as ordinary smallpox, but to have a milder course of disease. The infectiousness of people with modified smallpox would be 33 percent of that for people with ordinary smallpox, with a case fatality rate of 10 percent. However, it would be harder to recognize modified smallpox and cases would be slower to withdraw to the home or go to the hospital than for ordinary smallpox. Before smallpox is recognized in the hospital (i.e., the first period), 75 percent of cases would be recognized on the fourth day of rash and the remaining 25 percent on the seventh day of rash. After this, the smallpox is known to be present (i.e., the second period), and all ordinary smallpox would be recognized in the hospital on the third day of rash.

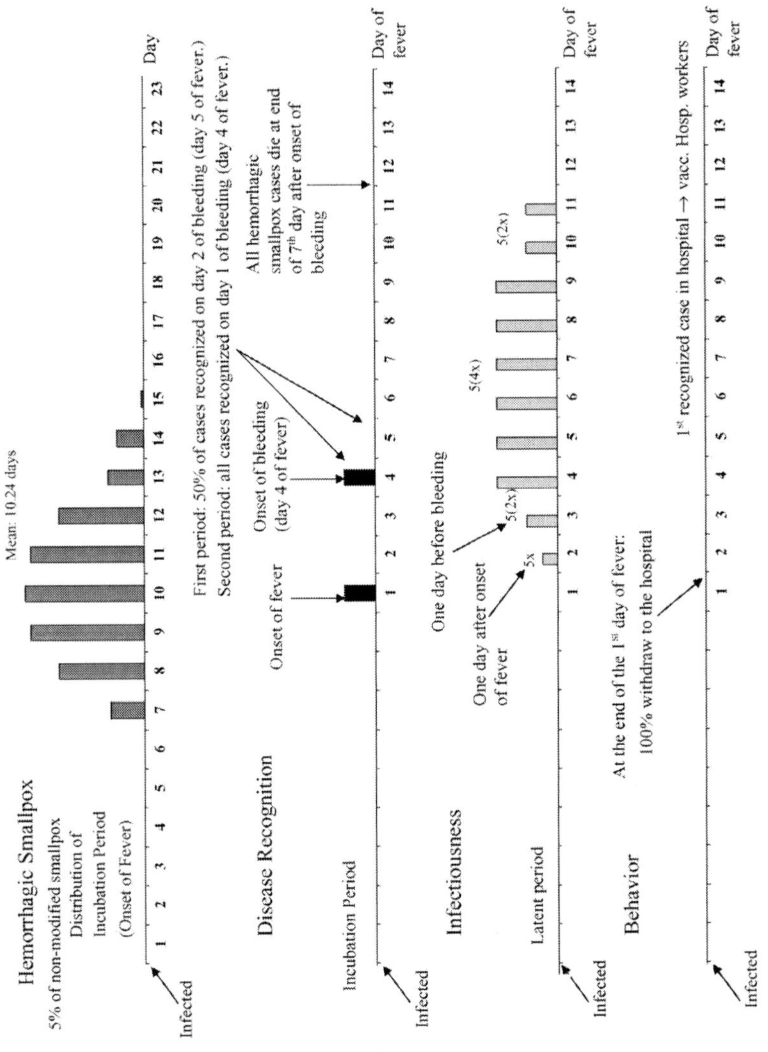

FIGURE 3-5 The natural history of hemorrhagic smallpox in terms of time lines. Hemorrhagic smallpox is assumed to have a shorter natural history and more severe disease progression than ordinary smallpox. Infected people would begin internal bleeding four days after the onset of fever, and 100 percent would die on the seventh day after the onset of bleeding. Before smallpox is recognized, we assumed that 50 percent of hemorrhagic smallpox cases would not be recognized and 50 percent would be recognized on the fifth day of fever. After smallpox is recognized, all hemorrhagic cases would be recognized on the fourth day of fever.

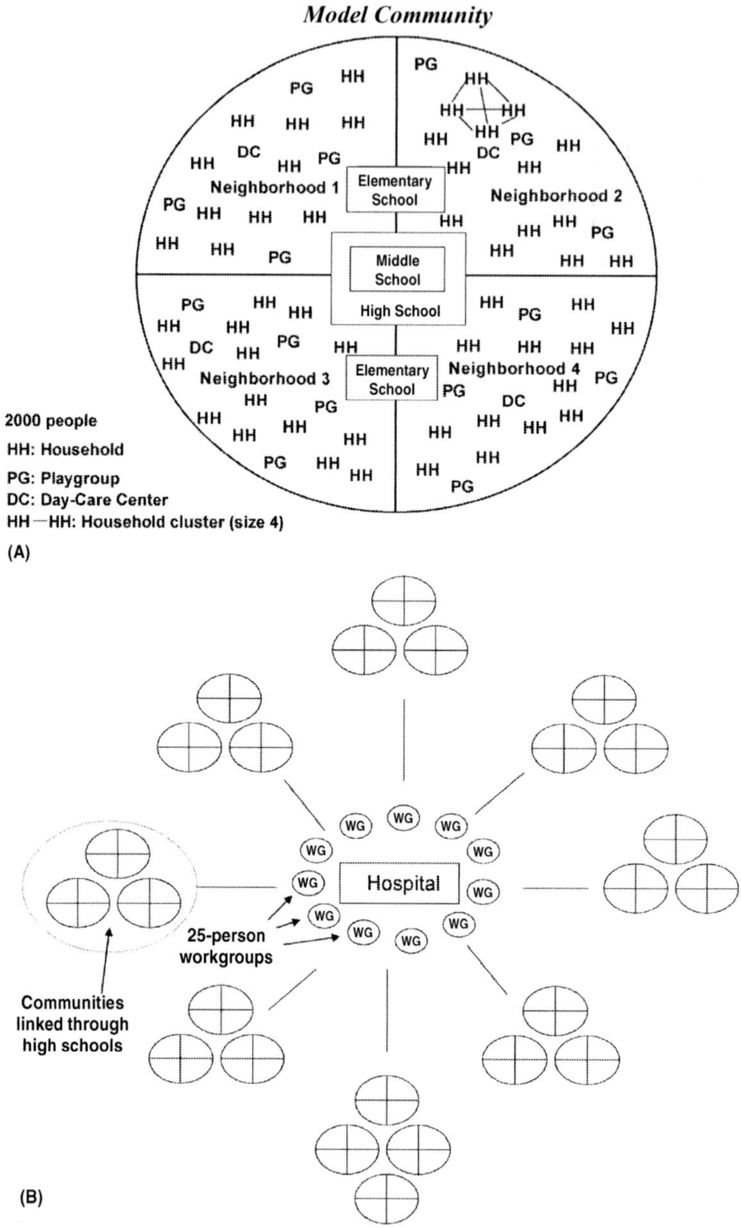

FIGURE 3-6 Structure of the populations. (A) The 2000 person subpopulations consist of households and household social clusters depicted by the connecting lines in neighborhood 2. Each subpopulation is partitioned into four neighborhoods. Small children mix

continued

sion probability would be 0.20 for days 4–9, and it would drop back down to 0.10 for days 10–14. People who complete the full course of disease without dying are considered to be immune.

Each day, for each susceptible, the probability of becoming infected is calculated based on his vaccination status, who is infectious in his or her mixing groups, and his or her vaccination status, as well as the mixing group-specific transmission probabilities. As an example, consider the simplest case that no one is vaccinated and we ignore the complex natural history of smallpox for illustrative purposes. An elementary school child is exposed to the number of child and adult infectives in his household I_{hc} and I_{ha}, his household cluster I_{kc} and I_{ka}, his school I_s, his neighborhood I_n, and the community I_m with corresponding transmission probabilities for each contact of x_{hcc} (child-to-child), x_{hac} (adult-to-child) . . . respectively. Then, symbolically, the probability P for that child to become infected on that day is:

$$P = 1 - (1 - x_{hcc})^{1hc} (1 - x_{hac})^{1ha} (1 - x_{kcc})^{1kc} (1 - x_{kac})^{1ka} \\ \times (1 - x_s)^{1s} (1 - x_n)^{1n} (1 - x_m)^{1m}$$

This equation is evaluated term-by-term in order to identify the source of infection if an infection occurs. Once infected, a person passes through the natural history of the infection process (Figures 3-3 through 3-5). The length of the incubation period is randomly selected from the probability distributions. The rest of the disease progression follows deterministically as indicated in the Figures. Aspects of the infected person's behavior, such as if and when he or she withdraws to the home or goes to the hospital, are simulated stochastically according to the probability distributions in Figures 3-3 through 3-5. The model assumes the same contact structure each day of the simulation with the exception of trips to the hospital.

We created a person-to-person graph of our population by constructing a contact structure proportional to the transmission probabilities given in Table 3-4 (see Annex 3-1). The resulting weighted graph has an average clustering coefficient of 0.48, much larger than the average clustering coefficient of 10^{-3} of an Erdös-Rényi random graph with the same number of vertices and average degree (Watts and Strogatz, 1998). The average shortest path was four people. This

---

in playgroups and daycare centers within their neighborhoods. The school mixing groups link neighborhoods as shown. (B) Clusters of the subpopulations are created by allowing ten percent of high school students in each of the clusters of subpopulations to mix with high schools in other subpopulations in the same cluster. All adults who work are randomly assigned to work in mixing groups of size 25 throughout the whole population. In addition, all people can attend a single hospital.

large clustering coefficient and small average shortest path suggest that we have a small-world person-to-person contact graph (Watts and Strogatz, 1998; Watts, 1999). This indicates that the smallpox transmission will tend to infect close-knit groups such as households, daycare centers, and schools when introduced, and then remain confined to these groups for a few generations of transmission. After this, transmission will tend to jump to somewhat more socially distant groups in a sporadic fashion. This pattern of local clustering followed by larger jumps, makes smallpox susceptible to perifocal control efforts, such as surveillance and containment. This is in contrast to the rapid uniform transmission that would occur if the contact structure of the person-to-person graph was more like a random graph, i.e., random mixing. In this case, perifocal control measures would probably tend to fail (Kaplan et al., 2002).

Interventions

For those people who receive a fresh smallpox vaccination before they are infected, we assumed the vaccine efficacy is 0.97, and that response to vaccination is all-or-none (Halloran et al., 1997). For those who receive a fresh vaccination four days post-infection, we assumed that 90 percent would not develop disease and 10 percent would develop modified smallpox. For those vaccinated between 5–7 days post-infection, 60 percent would develop modified smallpox, 38 percent ordinary smallpox, and 2 percent hemorrhagic smallpox. Vaccination reduces the death rate of breakthrough infections, for old vaccinations or fresh vaccinations 4–7 days post-infection, to a very low level, i.e., 1 percent or less.

We evaluated a number of intervention strategies. The most basic for traditional smallpox control has been surveillance and containment, also referred to as targeted or ring vaccination, which is part of the Centers for Disease Control and Prevention response plan (CDC, 2002). For this control strategy, when the first case of smallpox is recognized, all hospital workers who deal with smallpox cases would be immediately vaccinated. Recognized cases of smallpox would be placed in hospital-based isolation, and their close contacts would be vaccinated and kept under observation. These close contacts would be those people in the recognized case's household and, when appropriate, in the case's household social cluster, daycare center, school group, or workplace. Contacts in the neighborhood or the community at large would not be considered to be close contacts and not be isolated. Children under one year of age are not vaccinated.

We considered mass reactive vaccination where vaccination would begin one day after recognition of the first case of smallpox, and would take seven days to complete to a particular level. Smallpox cases are not vaccinated. In addition, any person freshly vaccinated through contact tracing or pre-emptive vaccination in the hospital would not be revaccinated. The schools would serve as vaccination centers and be closed for that seven-day period. A further strategy that was considered was the prevaccination of different proportions of hospital workers.

We also considered reactive closing of the schools for ten days, starting one day after recognition of the first case of smallpox.

In accordance with the working group, we evaluated a range of control scenarios (Table 3-5). The baseline scenario involved people withdrawing to the home and others being placed in effective hospital isolation at the appropriate times, but no contact tracing or vaccination. Scenario 3 was surveillance and containment (with vaccination of close contacts) alone, and then scenarios 4–10 involved surveillance and containment plus various additional control measures including pre-emptive vaccination of hospital workers, reactive mass vaccination, and reactive school closings. Scenarios 1 and 2 involved no interventions and were used to help validate the simulations. For the attack scenario we assumed that 500 randomly selected people are initially infected. For each intervention scenario, 100 epidemics were stochastically simulated.

Results

Calibration of the model was based on historical data available on smallpox, including household secondary attack rates (Fenner et al., 1988), relative age-specific attack rates being higher in children (Thomas et al., 1971), and the distribution of secondary cases produced by an introductory case (Fenner et al., 1988; Mack, 1972). We roughly calibrated the transmission probabilities in households to observed household secondary attack rates from past smallpox epidemics. These ranged from 44 percent to 88 percent to unvaccinated people in a variety of populations in Africa and South Asia in the 1960s and 1970s (Fenner et al., 1988; Mack, 1972). For example, we set the child-to-adult household daily transmission probability x to 0.05. Using the information in Figure 3-3, if the infected child remained in the household over his entire infectious period, then the probability that he would infect the exposed adult would be 0.85 (i.e., household secondary attack rate of 85 percent). However, in reality the household secondary attack rate would be lower as the infected child would be placed in isolation when recognized as a smallpox case. Thus, 85 percent is the maximum household secondary attack rate. If the index infected child had modified smallpox, then the maximum secondary attack rate for child-to-adult transmission in the household would be 46 percent, and if the index case had hemorrhagic smallpox, then the maximum secondary attack rate would be 100 percent. These relationships are shown in Figure 3-7. The maximum secondary attack rate for other mixing groups is illustrated on this plot.

Figure 3-8A shows the first 60 days of one stochastically simulated smallpox epidemic with 500 randomly selected initially infected people from all age groups in the population for surveillance and containment (scenario 3), while Figure 3-8B shows the same for surveillance and containment plus preemptive hospital worker vaccination at 50 percent and reactive school closing and mass vaccination at 80 percent (scenario 9). Figures 3-8A and B show that the model

TABLE 3-5 Smallpox Simulation Scenarios

	Scenario										
	1	2	Baseline	3	4	5	6	7	8	9	10
Background immunity		+[a]	+	+	+	+	+	+	+	+	+
Surveillance and containment			Isolation but no vaccination	+	+	+	+	+	+	+	+
Pre-emptive vaccination											
Pre-emptive vaccination (hospital only)					10%	50%	10%	50%	10%	50%	10%
Reactive school closure, 10 days							+	+	+	+	+
Mass reactive vaccination							40%	40%	80%	80%	

[a]The + indicates that the factor is present.

reproduces the characteristic epidemic waves of smallpox roughly every two weeks. Although an outbreak is not prevented, it is reduced to a low level.

Table 3-6 shows the number of cases, not counting the initial cases, when surveillance and containment, which includes vaccination of close contacts (scenario 3), is instituted. The total number of cases averages 828, with 50 percent of the cases from the hospital, 18 percent from the family or other close contacts, 19 percent from schools or the workplace, and 13 percent from the neighborhoods and community at large. This latter 13 percent of infecting contacts would be untraceable. These percentages are quite close to those observed for European smallpox epidemics for 1950–1971 (Table 3-7).

Table 3-8 gives the numbers of smallpox cases and deaths for the baseline and for scenarios 3–10. (Results for scenarios 1 and 2 are not in the Table 3-8 but are given below.) If the only action were the isolation of cases (baseline scenario), then the model predicts an average of 1,750 cases and 523 deaths. If we add vaccination and carry out surveillance and containment (scenario 3), then the average number of cases would be reduced to 828 and the number of deaths to 211. Figure 3-8A indicates that for surveillance and containment there would be a relatively large second wave of cases after the initial wave, and then a much smaller third

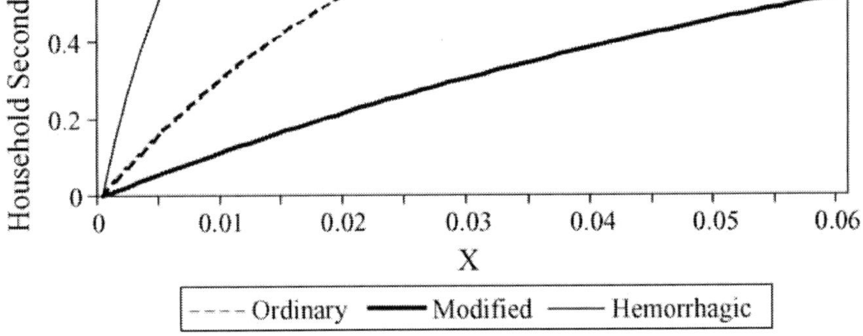

FIGURE 3-7 A plot showing the relationship between the transmission probability x during the second day of fever from an unvaccinated case of smallpox to an exposed unvaccinated person in a mixing group and the maximum household secondary attack (SAR) rate if he circulated in the mixing group for his entire infectious period. This relationship is based on the smallpox natural histories given in Figures 3-3 through 3-5.

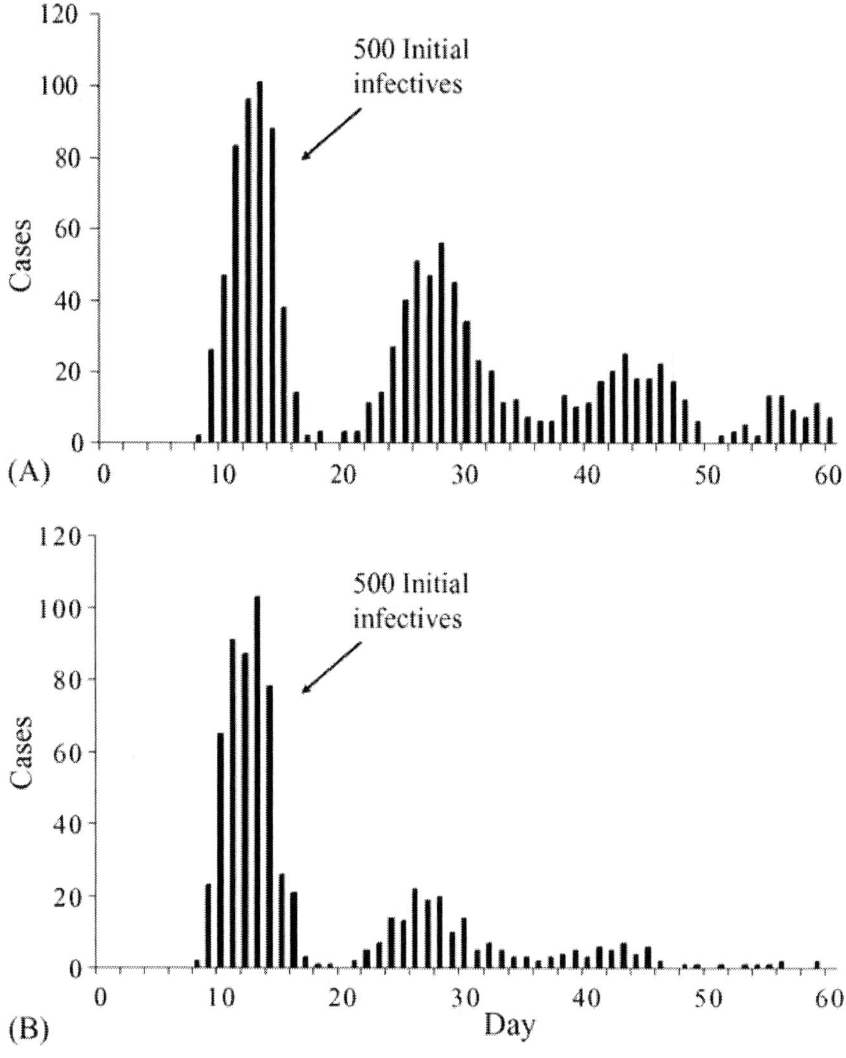

FIGURE 3-8 The first 60 days of one stochastically simulated smallpox epidemic with 500 randomly selected initially infected people from all age groups. (A) Epidemic surveillance and containment (scenario 3). This epidemic had a duration of 196 days, while the average duration of the epidemics under scenario 3 was 194 days. (B) Epidemic with surveillance and containment, 50 percent preemptive hospital vaccination and 80 percent reactive mass vaccination with reactive school closure for ten days (scenario 9). This epidemic had a duration of 91 days, while the average duration of epidemics under scenario 9 was 87 days.

TABLE 3-6 Distribution of Cases Excluding Initial Cases for Surveillance and Containment with Vaccination of Close Contacts (Scenario 3)[a]

	Cases		
Source of cases	Mean	Q1[b]	Q3[b]
Household	82	70	93
Neighborhood cluster	62	50	72
Daycare	6	3	8
Schools	101	78	120
Workgroup	60	19	87
Hospital (smallpox ward)	413	362	463
General neighborhood	52	44	60
Community	54	46	63
Total[c]	828	694	935

[a]Based on 100 simulations.
[b]Q1: first quartile; Q3: third quartile.
[c]A small number of people may have been infected from more than one source.

TABLE 3-7 Average Distribution of the Sources of Infections for Smallpox Cases with Surveillance and Containment (Scenario 3), Compared to the Distribution Observed in European Epidemics, 1950-1971

Location	Model	European
Hospital	50%	50%
Family or other close contact	18%	22%
Workplace or school	19%	14%
Unknown source	13%	14%

and fourth wave. Preemptive vaccination of 10 percent of hospital workers in addition to surveillance and containment (scenario 4) has a small effect on the average number of cases; however, preemptive vaccination of 50 percent of the hospital workers (scenario 5) has a relatively large effect on reducing the number of cases. Reactive mass vaccination of 40 percent of the susceptible population (scenarios 6 and 7) has a large additional effect. On average, 80 percent reactive mass vaccination (scenarios 8 and 9) is the most effective in reducing the outbreak to a minimal number of cases. By comparing scenarios 4 and 10, we see that reactive closing of the schools for ten days is not particularly effective.

Table 3-9 gives the numbers of fresh smallpox vaccinations for scenarios 3–10. Under surveillance and containment (scenario 3), an average of 7,501 fresh doses of vaccine would be used, far fewer than the ~25,500 doses that would be used under 40 percent reactive mass vaccination plus surveillance and containment and preemptive hospital vaccination (scenarios 6 and 7) or the

TABLE 3-8 Scenario Results, Excluding the 500 Initial Cases[a]

	Cases			Deaths		
Scenario: action	Mean	Q1[b]	Q3[b]	Mean	Q1	Q3
Baseline	1,750	1,527	1,428	523	455	584
3:SC[c]	828	694	935	211	173	239
4:SC + PHV10%	768	653	872	197	168	231
5: SC + PHV50%	678	595	750	180	156	206
6: SC + PHV10% + RSC + RMV40%	439	390	474	96	84	107
7: SC + PHV50% + RSC + RMV40%	367	341	394	83	73	91
8:SC + PHV10% + RSC + RMV80%	253	218	276	38	732	44
9: SC + PHV50% + RSC + RMV80%	203	185	219	33	29	38
10: SC + PHV10% + RSC	712	152	798	182	152	206

PHV, pre-emptive hospital vaccination; RSC, reactive school closure; RMV, reactive mass vaccination.

[a]based on 100 simulations.

[b]Q1: first quartile; Q3: third quartile.

[c]SC, surveillance and containment with vaccination of close contacts.

~45,000 doses that would be used under 80 percent reactive mass vaccination plus surveillance and containment and preemptive hospital vaccination (scenarios 8 and 9). The average number of doses used with surveillance and containment decreases with increasing level of preemptive hospital worker vaccination, since the total number of vaccinations due to contact tracing is decreased due to fewer cases. This can be seen by comparing the number of vaccine doses needed for scenarios 3–5.

For orientation purposes, we ran the simulator assuming no prior immunity, no interventions, and that cases do not withdraw to the home or go to the hospital (scenario 1). Nearly the entire population is infected, an average of 49,500 cases. This result is expected since infected people are modeled to circulate in the community over their entire infectious period. The average number of deaths is 16,598 people. The addition of prior immunity (scenario 2) makes a small difference in the number of cases, averaging 46,643 people. Prior immunity makes a larger difference in the number of deaths, an average of 13,681. This decrease is mostly due to the increased number of modified smallpox cases among those people previously vaccinated. In addition, by comparing the baseline average of 1,750 cases to the average of 46,643 cases under scenario 2, we see the great effectiveness of people with early smallpox symptoms simply withdrawing to the home and entering hospital isolation.

Table 3-10 shows the results of a sensitivity analysis where we vary the time it takes to recognize a case and begin isolation of the case and vaccination of close contacts under the surveillance and containment scenario 3 (see case recognition days in Figures 3-3 through 3-5). This further delay could be the result

TABLE 3-9 Number of Vaccine Doses[a]

Scenario	Doses Mean	Q1[b]	Q3[b]
3	7,501	6,825	7,966
4	7,221	6,542	7,772
5	6,725	6,231	7,185
6	25,677	25,481	25,856
7	25,472	25,267	25,667
8	45,246	45,203	45,284
9	45,214	45,178	45,262
10	6,888	6,336	7,357

[a]Based on 100 simulations.
[b]Q1: first quartile; Q3: third quartile.

of some cases not being caught quickly, confusion about smallpox symptoms, or some other problem with the medical response. If we delay an additional day, the number of cases and deaths doubles. More than one day further delay would result in a further approximately 50 percent increase in the number of cases and deaths.

Discussion

This work suggests that the current federal government policy of post-release surveillance and containment, if effectively implemented, could be sufficient to contain either a small or large intentional release of smallpox. We have shown that reactive mass vaccination in addition to surveillance and containment during an attack results in fewer cases and deaths than surveillance and containment alone. However, many more people would need to be vaccinated for reactive mass vaccination than for surveillance and containment. Since the risk of vaccine-related illness is about 10^{-4} and vaccine related death is about 10^{-6} (Henderson et al., 1999), one would expect an average of 2.5–4.5 vaccine-related illnesses and a small probability that one person would die due to vaccination for the reactive mass vaccination strategies considered here. If logistically possible, implementation of reactive mass vaccination would make sense. Prevaccination of hospital workers results in somewhat smaller outbreaks in the event of an attack. However, since it is not known when or where an attack may occur, prevaccination strategies would require that large numbers of people be vaccinated throughout the entire country. This is true of any prevaccination program before an attack. Such programs either for hospital workers and first responders or for the general population may not be necessary given the effectiveness of surveillance and containment that could be carried out at the location of an attack. The benefits of such prevaccination need to be weighed against the potential harm that would

TABLE 3-10 Surveillance and Containment for Various Delays in Case Recognition, Excluding the 500 Initial Cases[a]

	Cases			Deaths		
Additional delay in recognition (days)	Mean	Q1[b]	Q3[b]	Mean	Q1	Q3
Current model[c] (no additional delay)	828	694	935	211	173	239
1	1,681	1,509	1,848	416	370	459
2	2,017	1,879	2,162	503	461	533
3	2,217	1,995	2,373	578	522	625
4	2,372	2,081	2,601	658	585	720
5	2,786	2,574	3,007	780	720	841

[a]Based on 100 simulations.
[b]Q1: first quartile; Q3: third quartile.
[c]Model default: smallpox cases are recognized in the hospital either seven days (ordinary and modified cases) or four days (hemorrhagic cases) after onset of fever.

ensue due to vaccine-related injury. The quantitative validity of the above statements depends on the assumptions, parameter values, and model structure that we have used here. However, our general conclusions should be robust to this uncertainty.

Children under one year of age do not receive smallpox vaccine. However, in the absence of maternal antibodies, young children are at very high risk of serious disease and death if they contract smallpox. This makes the surveillance and containment policy very important for these children since the rapid vaccination of family members of index cases and of school children affords very young child indirect protection.

To assess the robustness of our conclusions about the effectiveness of control strategies modeled, we carried out a number of sensitivity analyses not given in the results. The total number of smallpox cases was found to be sensitive to variation in the transmission probabilities x in the different mixing groups. However, the relative effectiveness of the control strategies was not affected across the range from small to larger values of x. The most sensitive factor was timing of withdrawal to the home and isolation of cases. A delay in recognition of cases by one or more days beyond the hypothesized control strategy outlined in Figures 3-3 through 3-5 was found to result in poorly contained simulated epidemics (Table 3-10). The sensitivity analysis also reflects uncertainty about the exact onset of infectiousness relative to symptoms, since earlier than hypothesized onset of infectiousness would be equivalent to a delay in isolation. This result is consistent with a previous modeling exercise that showed logistical delays in fully implementing surveillance and containment could lead to a large outbreak (Kaplan et al., 2002).

We created a 50,000-person model population based on the U.S. census 2000 information and our conception of how a typical American population is connected in terms of potential smallpox transmission. To assess whether we have the approximate connectivity of a typical U.S. population, we compared our person-to-person graph to the graph that was constructed from individual level daily travel and location visited survey data from Portland, Oregon, with a population of 1.6 million people (Eubank et al., 2004). The average clustering coefficient for both graphs is 0.48. This indicates that the degree to which the two populations are clustered into close mixing groups such as households, schools, and workplaces is similar. The mean shortest path for the Portland population was six, while it is four for our population. Thus, the links between clusters for our population are somewhat shorter than those in Portland. Both our graph and the one for Portland are small world with similar characteristics. Although our population is smaller than the Portland population, the connectivity of any person with others in the population is roughly similar for the two populations. Thus, we believe that our simulation population of 50,000 people is large enough to investigate the effectiveness of the various containment strategies against a large attack.

Our previous modeling work has shown that surveillance and containment would be effective in containing and sometimes preventing a smallpox outbreak for a small number of initial cases (Halloran, 2002b). In this work, using a model with different epidemiologic parameter values, we show that surveillance and containment could be effective in containing an outbreak with a large number of initial cases. This suggests that further prevaccination of the population of the USA would be counter-productive. However, a rapid and well-organized response to a smallpox bioterrorist attack would be needed to make containment efficient.

Acknowledgment

This research was partially funded by the Fogarty International Center, National Institute of Allergy and Infectious Disease grant R01AI32042, and the National Institute of General Medical Sciences MIDAS grant U01GM070749. Conflict of interest: No conflict of interest to declare.

INDIVIDUAL-BASED COMPUTATIONAL MODELING OF SMALLPOX EPIDEMIC CONTROL STRATEGIES[38]

Donald S. Burke, M.D.[39]
Johns Hopkins University

Joshua M. Epstein, Ph.D.[40]
The Brookings Institution

Derek A. T. Cummings, Ph.D.[41]
Johns Hopkins University

Jon I. Parker, B.S.
The Brookings Institution

Kenneth C. Cline, B.S.
Johns Hopkins University

Ramesh M. Singa, B.S.
Johns Hopkins University

Shubha Chakravarty, M.S.
The Brookings Institution

Agent-based modeling is a new scientific approach in the tradition of cellular automata made possible by recent advances in computing. In brief, an agent-based model or "artificial society" includes some population of "agents," typically individual human beings, each implemented as a distinct object or data structure in a computer program. These agents interact locally with one another in the computer code. Over many iterations, these microinteractions can generate large-scale macroscopic phenomena of fundamental interest, specifically, the course of epidemics in space and time. When calibrated to actual epidemic data, these models become credible bases for policy analysis.

One distinguishing feature of individual-based models is that the individuals in the model (the people) are not aggregated into a few homogeneous pools (e.g., susceptibles, infectives, and removeds) (Longini et al., 2006). Rather, agent populations are highly heterogeneous; every single individual is explicitly modeled

---

[38]Reprinted from Burke DS, et al. 2006. Individual-based computational modeling of smallpox epidemic control strategies. *Academic Emergency Medicine* 13(11):1142-1149. Copyright 2006, with permission from Elsevier.

[39]Johns Hopkins Bloomberg School of Public Health, Department of International Health.

[40]Center on Social and Economic Dynamics.

[41]Department of International Health, The Bloomberg School of Public Health.

and tracked. A second distinguishing feature of agent models is that events transpire on an explicit space. In the agent approach, people interact only with neighbors in the space, who change as agents move around, for example, coming and going to school or work in the model. Our simulation approach attempts to create a realistic depiction of local social contact dynamics in discrete and explicitly represented social units (the home, the hospital, and so on). In agent models where individual decision making is present, agents have bounded rationality and typically make use of simple rules based on local information (Berry et al., 2002).

All of this action can be depicted graphically in real time, as if looking down on the social space from above, watching agents move to and from the various social units (homes, schools, workplaces), changing colors as they progress through the phases of the disease. It is centrally important to note that, because of their explicit inclusion of physical space, local interactions, and individual heterogeneity, the agent models produce fundamentally different spatiotemporal epidemic dynamics than smoothed differential equation models (Epstein, 1999). When this highly visual mode of modeling produces a novel hypothesis, all the graphics can be turned off, and the model can be run stochastically for millions of cycles to explore the robustness of the finding. There is no sacrifice of rigor; indeed, we can generate extremely high volumes of clean data and analyze it statistically to yield a very high-fidelity characterization of model behavior and robust calibration to historical data.

The modeling approach taken derives directly from that previously developed by Epstein et al. (2004). This work was undertaken under the auspices of the Smallpox Modeling Working Group of the U.S. Department of Health and Human Services Secretary's Advisory Council on Public Health Preparedness. The charge to members of the working group was to modify their existing smallpox epidemic models to incorporate agreed-on values for the natural history and transmission of the disease and then use the model to analyze epidemic outcomes under a variety of plausible attack and response scenarios.

Modeling

Assumptions and Parameters

This article describes an extended version of our basic agent-based smallpox model, distributed over a range of social architectures, population sizes, and assigned response scenarios (Epstein et al., 2004). Throughout this article, we use the detailed biomedical assumptions agreed on by the Smallpox Modeling Working Group.[42] Values assigned included the distribution of ordinary, modified, and

[42]Other members of the Smallpox Modeling Working Group of the Secretary's Advisory Council on Public Health Preparedness, U.S. Department of Health and Human Services, who served in advisory

hemorrhagic smallpox to be expected in the U.S. population and, for each form of the disease, the assumed distributions of incubation periods, time to disease recognition, infectiousness, and mobility effects of infection and disease (Longini, 2006). Other parameters, such as the probability of smallpox transmission per contact at various phases of the disease's natural history, and contact rates per day in various social units (the home, school, workplace, and hospital) were estimated by calibration of the model to real epidemiologic data. A detailed description of all aspects of the model is included in the technical appendix.[43]

Simulated Social Structures

The basic social structure represented in the model was the town. Towns were composed of smaller social units representing households, workplaces, schools, and hospitals. The basic model unit of time was the "day," which was parsed into two equal halves, a nighttime when all healthy individuals were at home and a daytime when healthy individuals were at work or school. No genders were assigned. Adults cycled between work and home, whereas children cycled between school and home.

The number of persons per household was distributed according to U.S. census figures (U.S. Census, 2000). For the "uniform" towns, all adults in a town went to one workplace, and all children went to one school. For the "hub-and-spoke" and "ring" towns (see following text), the town was divided into districts, each with its own local workplace and local school. Where the town architecture was composed of more than a single uniform district, 10 percent of workers "commuted" to the workplace of a contiguous district. Children attended school in the town in which they lived. For all towns, there was a single hospital, including 150 adult health-care workers who worked at the hospital rather than at the common workplace. During the daytime and nighttime, contacts were made with one of the eight possible physically neighboring individuals on the grid surface, wherever the individual was at that time (household, workplace, and so on). The proportion of contacts that were potential transmitting contacts in each social unit was assigned according to the calibration of the model to historical data (Mack, 1972).

In this study, we examined models for two sizes of total population: 6,000-person towns and 50,000-person towns. For the 6,000-person towns, we examined three social architectures: a single uniform town of 6,000 persons, a ring town

---

role to this work included the following: J Chin, M.D., M.P.H. (Working Group Chair; University of California, Berkeley), L Anderson, M.D. (CDC), L Borio, M.D. (HHS), J Breman, M.D., D.T.P.H. (Fogarty International Center, NIH), G Curlin, M.D. (National Institute of Allergy and Infectious Diseases, NIH), J Donlon, M.D., Ph.D. (HHS), E Eitzen, M.D., M.P.H. (HHS), JW Glasser, M.D. (CDC), ME Halloran, M.D., D.Sc. (Emory University), DA Henderson, M.D. (HHS), IM Longini, Ph.D. (Emory University), E McKenzie, Ph.D. (Fogarty International Center, NIH), M Miller, M.D. (Fogarty International Center, NIH), and F Murphy, D.V.M., Ph.D. (University of California, Davis).

[43]Available as a Data Supplement at http://www.aemj.org/cgi/content/full/j.aem.2006.07.017/DC1.

of six districts of 1,000 persons each, and a hub-and-spoke town with a single central district of 2,000 persons surrounded by four districts of 1,000 persons (Figure 3-9A and Figure 3-9B). In each town there was a single hospital that served all districts but was staffed by persons from one district. Index cases were always seeded into one district, the district containing the hospital. The social architectures for 50,000-person towns were analogous but larger.

Calibration of the Model to Historical Data

The parameters governing the probability of smallpox transmission per contact and the contact rates in different social settings were chosen through a calibration of simulated epidemics with historical data. The historical data describe outbreaks resulting from 49 importations of a single case of smallpox into nonendemic Europe during the period from 1950 to 1971 (Mack, 1972). Two distributions from these real epidemic data were used for the calibration: 1) the distribution of the total number of cases resulting from each of these importations and 2) the distribution of the location where transmissions occurred (in a hospital setting, in a workplace or school setting, or in the home). A parameter sweep was performed in which the per-contact transmission probability and the contact rate in the hospital, the home, and the workplace or school were systematically varied. The resultant distributions of epidemic sizes and transmission locations from 1,000 calibrating simulation runs were compared with the known historical distributions. The model parameter settings for per-contact transmission and contact rates in each social setting were chosen that minimized the sum of squared deviations from the known historical distributions.

Baseline "No Response" Scenarios

We first conducted an evaluation of simulated epidemics in the unrealistic setting of a complete absence of public health response. Two base case "no response" scenarios were examined. These scenarios served as baselines for comparison of intervention strategies. Scenario 1 presupposed only a low level of immunity among adults due to vaccination 30 years ago or more. Scenario 2 was identical to scenario 1 except that transmission from hospitalized persons was set at zero to determine if the force of infection in the community alone was sufficient to sustain the simulated epidemics.

Evaluation of "Response" Scenarios

We then conducted an analysis of eight response scenarios specified by the working group. These scenarios involved contact tracing and vaccination of family, coworker, and hospital contacts, hospital isolation of all cases, pre-emptive vaccination of health-care workers, school closures, and mass reactive

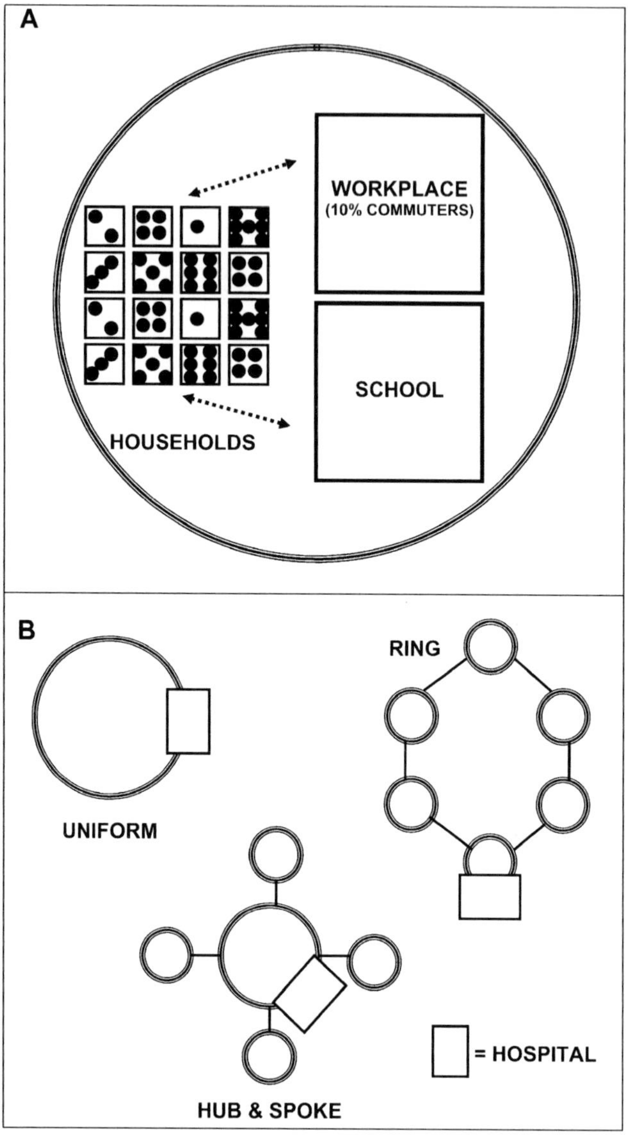

FIGURE 3-9 Schematic diagram of the social structures represented in the model. (A) Basic town/district social unit with multiple households, one school, and one workplace. (B) Town social architectures of single uniform, ring, and hub-and-spoke structure.

vaccination. The complete spectrum of scenarios is summarized in Figure 3-10. For each of the three architectures of the 6,000-person town model, each of the eight response scenarios was run 35 different times, with a different random seed each time. Note that the model is stochastic; the use of a different random seed for each run ensures that particular realizations of the same model produce different random contact patterns in the population. The random seeds also ensure that model parameters drawn from set distributions (see Longini et al., 2006 and Figure 3-9) vary from individual to individual and run to run. For each combination of town architecture and response scenario, we report here the mean and the standard deviation (over the 35 simulated epidemic runs) for the following key epidemic outcome measurements: total number of cases, total number of deaths, total number of vaccinations, and total epidemic duration.

For the 50,000-person town model runs, the analysis format was similar to that for the 6,000-person towns, except that here 500 initial infected individuals were introduced into the population. Because these runs absorb substantially more computational resources than the comparable 6,000-person town simulations, we explored only two architectures: the single uniform large town of 50,000 and a ring of six districts of equal size. Instead of 35 simulated epidemic runs per scenario-architecture pair, we present the statistics for just ten stochastic realizations.

Results

Tabular Results

Cases, deaths, vaccinations, and epidemic durations for simulated epidemics under the two "no response" and the eight response scenarios are shown in Tables 3-11 and 3-12 (see Annex 3-1) for populations of 6,000 individuals and 50,000 individuals, respectively.

Simulated Epidemics in the Baseline "No Response" Scenario

Simulations under the highly unrealistic "no response" scenarios 1 and 2 gave rise to large and lengthy epidemics. Each index case, on average, initiated an epidemic chain of transmission that subsequently infected hundreds of other individuals.

Evaluation of Response Scenarios in 6,000-Person Towns

Response scenarios 3–10 were all examined in model populations with each of the three (single uniform, ring, and hub-and-spoke) town architectures, and 35 simulated epidemics were run for each scenario in each architecture. Results are displayed in Table 3-11 (see Annex 3-1). Although there are some minor differences in the impact of the different response scenarios in different town struc-

	No Response Scenarios		Response Scenarios							
	1	2	3	4	5	6	7	8	9	10
Background immunity	Yes	→								
Hospital transmissions censored	Yes	Yes	No	→						
Household contact trace 100%			Yes	→						
Work/school contact trace 75%			Yes	→						
Hospital Vac & Isol (> 20 days)			Yes	→						
Preemptive Hospital Vaccination				10%	50%	10%	50%	10%	50%	10%
Reactive School Closure (10 days)						Yes	→			
Mass Reactive Vaccination						40%	40%	80%	80%	None

FIGURE 3-10 Summary sketch of the interventions and combinations of baseline conditions and interventions studied in "no response" scenarios 1 and 2 and response scenarios 3-10.

tures, the results for each response scenario show substantial concordance across all three town structures. Results are therefore presented in the text by scenario rather than by town structure.

In striking contrast to the "no response" scenarios 1 and 2, the interventions in scenario 3 alone are sufficient to limit the epidemic to a mean of fewer than 48 cases and a mean duration of less than 77 days. The addition of 10 percent or even 50 percent preemptive vaccination of hospital workers (scenarios 4 and 5) does not add appreciably to the protections afforded in scenario 3. Mass reactive vaccination of either 40 percent or 80 percent of the total population (scenarios 6–9) does provide some additional protection of the population, reduces the mean number of infected persons to fewer than 33, and shortens the mean epidemic duration to less than 60 days. The number of vaccinations given per infected person in scenarios 3, 4, and 5 was relatively low, averaging about 325 vaccinations per epidemic response. While reactive mass vaccination was effective in reducing the number of infections, the number of vaccinations per infected person was substantially greater under these scenarios (scenarios 6–9). School closure for ten days without mass vaccination (compare scenario 10 with scenario 4) appeared to provide little additional protection. The single uniform town tended to generate a higher mean total number of infected persons than did the more structured ring or hub-and-spoke town architectures, but this difference was not consistent across all scenarios.

Evaluation of Response Scenarios in 50,000-Person Towns

Response scenarios 3–10 were examined for 50,000-person towns with the single uniform and the ring architectures (but not the hub-and-spoke architecture); ten simulation runs were performed for each scenario for each town structure. Results of the epidemic simulations from the 50,000-person town are generally in close agreement with the results from the 6,000-person town presented previously. Specifically, under "no response" scenario 1, the epidemics run to near saturation with a large proportion of the population becoming infected. Epidemics under scenario 2 infect roughly half the population in both town sizes. The interventions in scenario 3 alone are sufficient to limit the epidemic to less than 10 percent of the population. Addition of 10 percent or 50 percent preemptive vaccination of hospital workers does not confer any appreciable additional protection. Mass reactive vaccination of either 40 percent or 80 percent of the population (scenarios 6–9) in the 50,000-person town model does provide some additional protection of the population, but the effect is less pronounced than in the 6,000-person town; this may reflect a higher proportion of persons initially infected in the 50,000-person town model (500, or 1 percent vs. 10, or 0.16 percent). School closure had no additional appreciable epidemic impact in the 50,000-person town model (scenario 10 vs. scenario 4).

Mass Vaccination

Four of the response scenarios examined in this project involved reactive mass vaccination, with population vaccine coverage of 40 percent (scenarios 6 and 7) or 80 percent (scenarios 8 and 9). Although these scenarios yielded smaller and shorter epidemics than response scenarios based on isolation, contact tracing, and targeted vaccination alone, the largest difference attributable to mass vaccination in the 6,000-person town was a reduction from 45.9 to 17.3 cases (scenario 4 [no mass vaccination] vs. scenario 9 [80 percent vaccine coverage]). By graphing the total epidemic size versus the number of vaccinations in each scenario, we calculate that the marginal "vaccine cost" to prevent one additional case is 172–213 vaccinations in the 6,000-person town/10-case attack model and 33–39 vaccinations in the 50,000-person town/500-case attack model.

Unique Features of the Model

To appreciate the unique features of this intrinsically stochastic agent-based modeling approach, a detailed analysis of data output from one scenario (scenario 3) in one population size (6,000-person town) with one town architecture (ring) is presented. Thirty-five epidemic simulation runs of this particular configuration were examined.

Epidemic size and duration. The size of epidemics ranged from a minimum of 12 cases (in addition to the ten attack cases) to a maximum of 83 cases, and epidemic durations ranged from a minimum of one epidemic generation beyond the attack (G_0) cases to a maximum of six epidemic generations. The mean (±SD) simulated epidemic duration was 3.9 (±1.2) epidemic generations after the initial generation (G_0) of 10 attack cases. On average, the 10-attack G_0 cases represented 23 percent of the epidemic case total. The first generation (G_1) wave of secondary cases alone accounted for 43 percent of the epidemic case totals; incidence thereafter declined rapidly with deployment of intervention strategies.

Epidemic reproductive rates. The epidemic reproductive rate (R) for each generation of cases ($RG_{i \to i+1}$) in each epidemic simulation run was calculated as the number of cases in each generation divided by the number of cases in the preceding generation (e.g., $RG_{0 \to 1}$ = cases G_1/cases G_0). This measure is not the same as a true R_0 because in the scenarios evaluated here multiple cases were introduced into the population and epidemic control strategies were activated on diagnosis of the first attack (G_0) case, thus minimizing transmission by G_0 attack cases that clinically progressed more slowly and transmitted later. The mean measured for $RG_{0 \to 1}$ (across 35 simulated epidemics) was 1.88. Subsequent mean generational reproductive rates, while epidemic control interventions were deployed in the model, were measured as $RG_{1 \to 2} = G_2/G_1 = 0.46$, $RG_{2 \to 3} = 0.43$, $RG_{3 \to 4} = 0.42$,

$RG_{4\rightarrow5}$ = 0.36, and $RG_{5\rightarrow6}$ = 0.35; none of the 35 simulated epidemics in the 6,000-person ring town scenario 3 lasted longer than six generations after the attack generation.

Clinical disease expression and transmission. In accordance with the values proposed by the working group expert advisors, cases were assigned one of three different clinical disease expressions: ordinary smallpox, modified smallpox, or hemorrhagic smallpox. Among all of the smallpox cases that occurred in these 35 simulated epidemic runs, 57 percent were ordinary, 40 percent were modified, and 3 percent were hemorrhagic cases. However, these proportions varied according to the epidemic generation (Figure 3-11A). In G_0, the proportions were 76 percent ordinary, 20 percent modified, and 5 percent hemorrhagic. In G_1 and subsequent generations, the proportions were 46 percent ordinary, 51 percent modified, and 3 percent hemorrhagic. These changes in clinical disease expression by epidemic generation reflect the effects of post-exposure vaccination of contacts, which rendered a number of cases to become modified that would otherwise have been ordinary cases (Fenner et al., 1988). We also calculated the epidemic reproductive rate for each clinical disease type for each epidemic generation. The initial disease-type epidemic reproductive rates, or $RG(\text{ordinary})_{0\rightarrow1}$, $RG(\text{modified})_{0\rightarrow1}$, and $RG(\text{hemorrhagic})_{0\rightarrow1}$, were respectively measured to be 1.70, 1.57, and 6.19. The overall initial epidemic reproductive rate $RG(\text{all cases})_{0\rightarrow1}$ value of 1.88 is a composite of these values. Subsequent epidemic generational values of R for ordinary, modified, and hemorrhagic smallpox were also calculated and are displayed in Figure 3-11B. After G_0, epidemic generational values of RG for modified smallpox were consistently higher than for other types of smallpox at the same generation of the epidemic.

Sensitivity of results to day of withdrawal. We examined the sensitivity of our results to a number of our model assumptions. Most notably, we found that our results were very sensitive to the assumption of the period of time that infected individuals who did not go to the hospital or withdraw to their home on the second day of fever would circulate in the community. These individuals were a small proportion of cases that were not ill enough to withdraw from the community until later in their course of illness. The base model assumes that all individuals will have withdrawn from the community (to the hospital) on day 4 of fever. In our sensitivity analysis, we varied this day from the day fever begins to day 5 of fever. This model change is roughly equivalent to varying the proportion of transmission that occurs before and after symptoms begin, a factor that other investigators have suggested is very important for the controllability of an infectious disease (Fraser et al., 2004). The number of cases that resulted in our model of 6,000 persons under scenario 2 varied dramatically as a function of this parameter from a mean of 2,981 cases in our base model assumptions to a mean of 174 cases if these individuals withdrew from the community on the day their fever began.

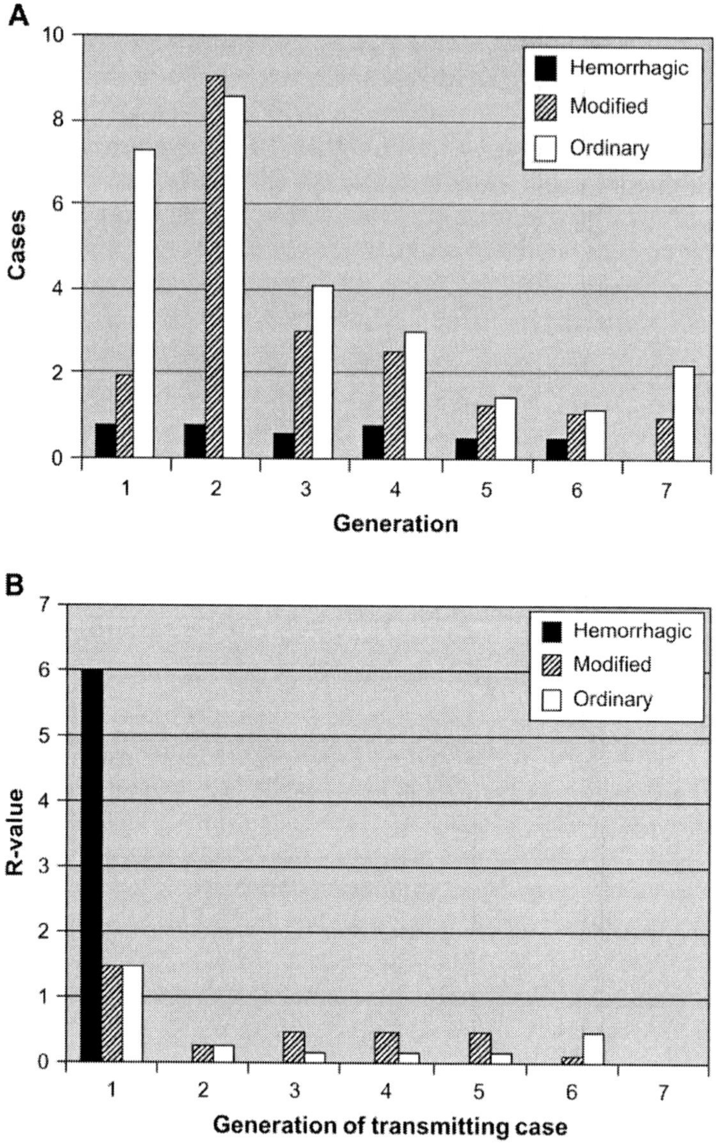

FIGURE 3-11 Number of cases of each clinical disease expression type and epidemic reproductive rate attributable to each of these clinical disease types for each epidemic generation. Values shown are averages from 35 epidemic simulations for the 6,000-person ring town under response scenario 3. Generation 0 = index (attack) cases. Ordinary (open bars), modified (striped bars), and hemorrhagic (closed bars) smallpox cases are shown. (A) Cases of each clinical disease type by epidemic generation. (B) Epidemic reproductive rate for each clinical disease type at each epidemic generation.

Discussion

The objective of this modeling exercise was to evaluate the potential effectiveness of epidemic control strategies that might be deployed in response to a bioterrorist attack. Our main finding is that contact tracing and vaccination of household, workplace, and school contacts, along with effective isolation of diagnosed cases, can control epidemics of smallpox. In our 6,000-person town model, we found that in scenario 3 (the combination of interventions most closely parallel to current U.S. governmental policies) the expected total number of smallpox cases that would ensue from ten simultaneous introductions would be 25–40 additional cases (CDC, 2002). Our findings in the 50,000-person town model were consistent with these estimates; under scenario 3,500 introductions into a population of 50,000 would give rise to approximately 1,100 new cases of smallpox. In both size versions of the model, reactive mass vaccination at the town level had additional value in bringing an epidemic under control. We estimate the number of reactive mass vaccinations required to incrementally reduce the epidemic by one case to be about 190 vaccinations in the 6,000-person town/10-attack-case model versions and about 35 vaccinations in the more intense 50,000-person town/500-attack-case model version.

Although a good deal of variation in the size and other characteristics of the modeled epidemics is expected in a highly stochastic epidemic model, we were nonetheless surprised by some of our observations (Bailey, 1953; Whittle, 1955). In our epidemic simulation runs, 1) epidemics ranged dramatically in size and duration based on chance alone, 2) the epidemic impact of individual index (attack) cases ranged from no transmissions whatsoever to large and lengthy transmission chains, and 3) the epidemic reproductive rate varied substantially by clinical disease type and by epidemic generation and was dependent on the underlying social network configuration. These results suggest that the heterogeneity of our microscale, agent-based model has significantly impacted the resultant epidemics.

Limitations

It is possible that some important parameters may not have been considered in the development of this model. For example, age-specific differences in the pathogenicity and transmissibility of smallpox were not considered, other than as they relate to age older than 32 years and prior vaccination status as well as social contact processes (schools for children vs. workplace for adults). We did not explicitly include risks of smallpox vaccination as a source of adverse outcome in our model. The number of vaccinations used in each modeled response is given in the Results section and can be used to estimate adverse outcomes. Another potentially important biological variable unexamined in this exercise is the effect of seasonality on transmission of smallpox (Fenner et al., 1988).

Perhaps the most important model parameters incompletely considered in this work are the social networks and contact processes that dictate disease transmission patterns. Clearly, there is a trade-off between the inclusion of a large degree of detail and heterogeneity in the social structure in a model and the complexity of the resultant model (Ferguson et al., 2003). We have included the level of social detail that we believed necessary to capture the transmission dynamics of smallpox. Although we explicitly modeled person-to-person contacts in hospitals, households, schools, and workplaces, our representations of these social units were admittedly crude. Although we addressed a range of model parameterizations and model structures, a larger sensitivity analysis may reveal surprising results. In future work, we will continue to examine the sensitivity of our results to specific model parameters.

Another limitation of this work is not the model itself but its proper interpretation and use. We caution that the numbers of cases generated in various scenarios should not be taken as quantitative predictions, but instead as a basis for comparing and evaluating different intervention strategies. We also note that in this exercise we modeled only a single geographically confined attack on a relatively small discrete social unit (6,000- or 50,000-person town). In the event of a real smallpox attack, response strategies would have to consider larger social networks and possible repeated introductions over a wide geographic area.

Conclusions

Our simulation exercise revealed that contact tracing and vaccination of household, workplace, and school contacts, along with prompt reactive vaccination of hospital workers and isolation of diagnosed cases, could contain smallpox at both epidemic scales examined. Individual-based simulations of smallpox epidemics provide a valuable tool in crafting policy regarding outbreak response.

REFERENCES

Annas G. 2002. Bioterrorism, public health, and civil liberties. *New England Journal of Medicine* 346(17):1337-1342.

APHA (American Public Health Association). 1918. Influenza: A report of the American Public Health Association. *Journal of the American Medical Association* 71:2068.

Bailey NTJ. 1953. The total size of stochastic epidemic. *Biometrika* 40:177.

Bayer R, Fairchild A. 2002. The limits of privacy: Surveillance and the control of disease. *Health Care Analysis* 10(1):19-35.

Bell DM, World Health Organization Working Group on International and Community Transmission of SARS. 2004. Public health interventions and SARS spread. *Emerging Infectious Diseases* 10(11):1900-1906.

Benenson AS, Ed. 1995. *Control of Communicable Diseases Manual*. Washington, DC: American Public Health Association.

Berry BJL, Kiel LD, Elliott E. 2002. Adaptive agents, intelligence, and emergent human organization: Capturing complexity through agent-based modeling. *Proceedings of the National Academy of Sciences* 99(3):7187-7188.

Bozzette SA, Boer R, Bhatnagar V, Brower JL, Keeler EB, Morton SC, Stoto MA. 2003. A model for a smallpox-vaccination policy. *New England Journal of Medicine* 348(5):416-425.

Burke DS, Epstein JM, Cummings DT, Parker JI, Cline KC, Singa RM, Chakravarty S. 2006. Individual-based computational modeling of smallpox epidemic control strategies. *Academic Emergency Medicine* 13(11):1142-1149.

CDC (Centers for Disease Control and Prevention). 2002. *Smallpox Response Plan and Guidelines* (version 3.0). [Online]. Available: http://www.bt.cdc.gov/agent/smallpox/response-plan/ [accessed February 14, 2007].

CDC. 2005a. *Public Health Guidance for Community-Level Preparedness and Response to Severe Acute Respiratory Syndrome (SARS)*. [Online]. Available: http://www.cdc.gov/ncidod/sars/guidance/I/index.htm [accessed January 26, 2007].

CDC. 2005b. *Regulatory Impact Analysis of Proposed 42 C.F.R. Part 70 and 42 C.F.R. Part 71*. [Online]. Available: http://www.cdc.gov/ncidod/dq/nprm/docs/draft_ria_final.pdf [accessed January 26, 2007].

CDC. 2005c. *Smallpox Vaccination Program Status by State*. [Online]. Available: http://www.cdc.gov/od/oc/media/pressrel/smallpox/spvaccin.htm [accessed February 15, 2007].

CDC. 2006. *Questions and Answers on the Executive Order Adding Potentially Pandemic Influenza Viruses to the List of Quarantinable Diseases*. [Online]. Available: http://www.cdc.gov/ncidod/dq/qa_influenza_amendment_to_eo_13295.htm [accessed April 12, 2007].

CDC. 2007. *Ethical Guidelines in Pandemic Influenza*. Advisory Committee to the Director, Ethics Subcommittee. [Online]. Available: http://www.cdc.gov/od/science/phec/panFlu_Ethic_Guidelines.pdf [accessed April 13, 2007].

Center for Law and the Public's Health. 2001 (December 21). *The Model State Emergency Health Powers Act*. Prepared by the Center for Law and the Public's Health at Georgetown and Johns Hopkins Universities, Washington, DC. [Online]. Available: http://www.publichealthlaw.net/MSEHPA/MSEHPA2.pdf; www.publichealthlawnet/msehpa2.pdf [accessed December 27, 2006].

Cetron M. 2006. *Community-Wide Strategies in Pandemic Planning*. Presentation delivered at Institute of Medicine workshop entitled "Modeling Community Containment of an Influenza Pandemic," Washington, DC.

Cetron M, Landwirth J. 2005. Public health and ethical considerations in planning for quarantine. *Yale Journal of Biology and Medicine* 78(5):329-334. Also reprinted on page 99 of this report.

Cetron M, Simone P. 2004. Battling 21st-century scourges with a 14th-century toolbox. *Emerging Infectious Diseases* 10(11). [Online]. Available: http://www.cdc.gov/ncidod/eid/vol10no11/04-0797_12.htm [accessed December 27, 2006].

Chen H, Smith GJ, Li KS, Wang J, Fan XH, Rayner JM, Vijaykrishna D, Zhang JX, Zhang LJ, Guo CT, Cheung CL, Xu KM, Duan L, Huang K, Qin K, Leung YH, Wu WL, Lu HR, Chen Y, Xia NS, Naipospos TS, Yuen KY, Hassan SS, Bahri S, Nguyen TD, Webster RG, Peiris JS, Guan Y. 2006. Establishment of multiple sublineages of H5N1 influenza virus in Asia: Implications for pandemic control. *Proceedings of the National Academy of Sciences* 103(8):2845-2850.

Cohen J, Enserink M. 2002. Rough-and-tumble behind Bush's smallpox policy. *Science* 298(5602):2312-2316.

Council on Ethical and Judicial Affairs. 2005 (January). *The Use of Quarantine and Isolation in Public Health Interventions*. Report of the Council on Ethical and Judicial Affairs. CEJA Report 1-I-05. [Online]. Available: http://www.ama-assn.org/ama1/pub/upload/mm/369/ceja_recs_1i05.pdf; www.amaassn.org/ama1/pub/upload/mm/31/quarantine15726.pdf [accessed December 27, 2006].

Eichner M. 2003a. Case isolation and contact tracing can prevent the spread of smallpox. *American Journal of Epidemiology* 158(2):118-128.

Eichner M. 2003b. Analysis of historical data suggests long-lasting protective effects of smallpox vaccination. *American Journal of Epidemiology* 158(8):717-723.

Eichner M, Dietz K. 2003. Transmission potential of smallpox: Estimates based on detailed data from an outbreak. *American Journal of Epidemiology* 158(2):110-117.

Enserink M. 2003. Infectious diseases. Smallpox vaccination campaign in the doldrums. *Science* 300(5621):880-881.

Epstein JM. 1999. Agent-based computational models and generative social science. *Complexity* 4(5):41-60.

Epstein JM, Cummings DAT, Chakravarty S, Singha RM, Burke DS. 2004. *Toward a Containment Strategy for Smallpox Bioterror: An Individual-Based Computational Approach*. Washington, DC: Brookings Institution Press.

Eubank S, Guclu H, Kumar VS, Marathe MV, Srinivasan A, Toroczkai Z, Wang N. 2004. Modelling disease outbreaks in realistic urban social networks. *Nature* 429(6988):180-184.

Fenner F, Henderson DA, Arita I, Jezek Z, Ladnyi ID. 1988. *Smallpox and Its Eradication*. Geneva, Switzerland: World Health Organization.

Ferguson NM, Keeling MJ, Edmunds WJ, Gani R, Grenfell BT, Anderson RM, Leach S. 2003. Planning for smallpox outbreaks. *Nature* 425(6959):681-685.

Fidler DP, Cetron MS. 2007. International Considerations. In Goodman RA, Hoffman RE, Lopez W, Matthews GW, Rothstein MA, Foster KL. 2007. *Law in Public Health Practice, Second Edition*. New York: Oxford University Press. Pp. 168-195.

Fraser C, Riley S, Anderson RM, Ferguson NM. 2004. Factors that make an infectious disease outbreak controllable. *Proceedings of the National Academy of Sciences* 101(16):6146-6151.

Garrett L. 2005 (July/August). The next pandemic? *Foreign Affairs* 84:3.

Germann TC, Kadau K, Longini IM Jr., Macken CA. 2006. Mitigation strategies for pandemic influenza in the United States. *Proceedings of the National Academy of Sciences* 103(15):5935-5940.

Global Security. 2005. *Flu Pandemic Mitigation: Quarantine and Isolation*. [Online]. Available:www.globalsecurity.org/security/ops/hsc-scen-3_flu-pandemic-quarantine.htm [accessed January 26, 2007].

Global Security. 2006. *Flu Pandemic Mitigation: Deaths*. [Online]. Available: www.globalsecurity.org/security/ops/hsc-scen-3_flu-pandemic-deaths.htm [accessed January 26, 2007].

Goodman RA, Moulton AD, Matthews G, Shaw F, Kocher P, Mensah G, Zaza S, Besser R. 2006. Law and public health at CDC. *Morbidity and Mortality Weekly Report* 55(Sup02):29-33. [Online]. Available: http://www.cdc.gov/mmwr/preview/mmwrhtml/su5502a11.htm [accessed January 29, 2007].

Gostin LO. 2007. *Public Health Law: Power, Duty, Restraint*. Second Edition. New York and Berkeley: Milbank Memorial Fund and University of California Press.

Gostin LO, Powers M. 2006. What does justice require for the public's health? Public health ethics and policy imperatives of social justice. *Health Affairs* 25(4):1053-1060.

Gronvall GK, Borio LL. 2006. Removing barriers to global pandemic influenza vaccination. *Biosecurity and Bioterrorism* 4(2):168-175.

Halloran ME, Struchiner CJ, Longini IM. 1997. Study designs for evaluating different efficacy and effectiveness aspects of vaccines. *American Journal of Epidemiology* 146(10):789-803.

Halloran ME, Longini IM, Cowart DM, Nizam A. 2002a. Community trials of vaccination and the epidemic prevention potential. *Vaccine* 20(27):3254-3262.

Halloran ME, Longini IM, Nizam A, Yang Y. 2002b. Containing bioterrorist smallpox. *Science* 298(5597):1428-1432.

Hammarlund E, Lewis MW, Hansen SG, Strelow LI, Nelson JA, Sexton GJ, Hanifin JM, Slifka MK. 2003. Duration of antiviral immunity after smallpox vaccination. *Nature Medicine* 9(9):1131-1137.

Hawryluck L, Gold WL, Robinson S, Pogorski S, Galea S, Styra R. 2004. SARS control and psychological effects of quarantine. *Emerging Infectious Diseases* 10(7):1206-1212.

Henderson DA, Inglesby TV, Bartlett JG, Ascher MS, Eitzen E, Jahrling PB, Hauer J, Layton M, McDade J, Osterholm MT, O'Toole T, Parker G, Perl T, Russell PK, Tonat K. 1999. Smallpox as a biological weapon: Medical and public health management. *Journal of the American Medical Association* 281(22):2127-2137.

Heyman D. 2005. Model operational guidelines for disease exposure control. *Center for Strategic and International Studies (CSIS)*. Pre-publication draft.

Heymann A, Chodick G, Reichman B, Kokia E, Laufer J. 2004. Influence of school closure on the incidence of viral respiratory diseases among children and on health care utilization. *Pediatric Infectious Disease Journal* 23(7):675-677.

HHS (Department of Health and Human Services). 2005a. *HHS Pandemic Influenza Plan*. [Online]. Available: http://www.hhs.gov/pandemicflu/plan/pdf/HHSPandemicInfluenzaPlan.pdf [accessed January 26, 2007].

HHS. 2005b. *Control of Communicable Diseases (Proposed Rule)*. 42 C.F.R. Parts 70 and 71.

HHS. 2006a. *HHS Pandemic Influenza Plan*. Washington, DC: U.S. Department of Health and Human Services. [Online]. Available: http://www.hhs.gov/pandemicflu/plan [accessed December 27, 2006].

HHS. 2006b. *HHS Pandemic Influenza Plan, Appendix 4. Principles of Modern Quarantine*. Washington, DC: U.S. Department of Health and Human Services. [Online]. Available: http://www.hhs.gov/pandemicflu/plan/ [accessed December 27, 2006].

Iton AB. 2006. Rationing influenza vaccine: Legal strategies and considerations for local health officials. *Journal of Public Health Management Practice* 12(4):349-355.

IOM (Institute of Medicine). 2005. *Quarantine Stations at Ports of Entry Protecting the Public's Health*. Washington, DC: The National Academies Press.

IOM. 2006. *Modeling Community Containment for Pandemic Influenza: A Letter Report*. Washington, DC: The National Academies Press.

Jackson MM, Lynch P. 1985. Isolation practices: A historical perspective. *American Journal of Infection Control* 13(1):21-31.

Johnson NP, Mueller J. 2002. Updating the accounts: Global mortality of the 1918-1920 "Spanish" influenza pandemic. *Bulletin of the History of Medicine* 76(1):105-115.

Kaplan EH, Craft DL, Wein LM. 2002. Emergency response to a smallpox attack: The case for mass vaccination. *Proceedings of the National Academy of Sciences* 99(16):10935-10940.

Kass NE. 2001. An ethics framework for public health. *American Journal of Public Health* 91(11):1776-1782. [Online]. Available: http://www.ajph.org/cgi/reprint/91/11/1776 [accessed December 27, 2006].

Kayman H, Oblorh-Odjidja A. 2006. Revisiting public health preparedness: Incorporating principles of social justice into pandemic preparedness planning for influenza. *Journal of Public Health Management Practice* 12(4):373-380.

Kobasa D, Jones SM, Shinya K, Kash JC, Copps J, Ebihara H, Hatta Y, Kim JH, Halfmann P, Hatta M, Feldmann F, Alimonti JB, Fernando L, Li Y, Katze MG, Feldmann H, Kawaoka Y. 2007. Aberrant innate immune response in lethal infection of macaques with the 1918 influenza virus. *Nature* 445(7125):319-323.

Kotalik J. 2005. Preparing for an influenza pandemic: Ethical issues. *Bioethics* 19(4):422-431.

Longini IM, Nizam A, Xu S, Unqchusak K, Hanshaoworakul W, Cummings DA, Halloran ME. 2005. Containing pandemic influenza at the source. *Science* 309(5737):1083-1087.

Longini IM, Halloran M.E. Nizam A, Yang Y, Xu S, Burke DS, Cummings DA, Epstein JM. 2006. Containing a large bioterrorist smallpox attack: A computer simulation approach. *International Journal of Infectious Diseases* 11(2):98-108.

Mack TM. 1972. Smallpox in Europe, 1950-1971. *Journal of Infectious Diseases* 125(2):161-169.

Mariner W. 2001. Bioterrorism Act: The wrong response. *National Law Journal* 24:A21.

Markel H, Stern AM, Navarro JA, Michalsen JR. 2006. *A Historical Assessment of Nonpharmaceutical Disease Containment Strategies Employed by Selected U.S. Communities During the Second Wave of the 1918-1920 Influenza Pandemic*. [Online]. Available: http://www.med.umich.edu/medschool/chm/influenza/assets/dtra_final_influenza_report.pdf [accessed on January 26, 2007].

Markovits D. 2005. Quarantines and distributive justice. *Journal of Law, Medicine and Ethics* 33(2):323-344.

Moulton AD, Gottfried RN, Goodman RA, Murphy AM, Rawson RD. 2003. What is public health legal preparedness? *Journal of Law, Medicine, and Ethics* 31(4):672-683. [Online]. Available: http://www2.cdc.gov/phlp/docs/moultonarticle.pdf [accessed January 29, 2007].

Murray CJ, Lopez AD, Chin B, Feehan D, Hill KH. 2006. Estimation of potential global pandemic influenza mortality on the basis of vital registry data from the 1918-20 pandemic: A quantitative analysis. *Lancet* 368(9554):2211-2218.

Normile D. 2006. Evidence points to migratory birds in H5N1 spread. *Science* 311(5765):1225.

Ontario Ministry of Health and Long-Term Care. 2005. *Ontario Health Plan for an Influenza Pandemic*. [Online]. Available: http://www.health.gov.on.ca/english/providers/program/emu/pan_flu/pan_flu_plan.html [accessed December 27, 2006].

Osterholm MT. 2005(July/August). Preparing for the next pandemic. *Foreign Affairs* 84:24-26.

Public Health Leadership Society. 2002. *Principles of the Ethical Practice of Public Health*. New Orleans: Public Health Leadership Society. [Online]. Available: http://www.apha.org/codeofethics/ethicsbrochure.pdf [accessed December 27, 2006].

Reich DS. 2003. Modernizing local responses to public health emergencies: Bioterrorism, epidemics, and the Model State Emergency Health Powers Act. *Journal of Contemporary Health Law and Policy* 19(2):379-414.

Rosenthal E. 2006 (April 15). Bird flu virus may be spread by smuggling. *New York Times*.

Rosling L, Rosling M. 2003. Pneumonia causes panic in Guangdong province. *British Medical Journal* 326(7386):416.

Rothstein M, Alcalde MG, Elster NR, Majumder MA, Palmer LI, Stone TH, Hoffman RE. 2003. *Quarantine and Isolation: Lessons Learned from SARS*. [Online]. Available: http://biotech.law.lsu.edu/blaw/cdc/SARS_REPORT.pdf [accessed January 26, 2007].

SARS Commission. 2006. vol. 1 and 2. *Spring of Fear*. [Online]. Available: http://www.sarscommission.ca/report/index.html [accessed January 26, 2007].

Shapiro v. Thompson. 1999. 394 US 618 (1969).

Smith CB, Battin MP, Jacobson JA, Francis LP, Botkin JR, Asplund EP, Domek GJ, Hawkins G. 2004. Are there characteristics of infectious diseases that raise special ethical issues? *Developing World Bioethics* 4(1):1-16. [Online]. Available: http://www.blackwell-synergy.com/doi/pdf/10.1111/j.1471- 8731.2004.00064.x?cookieSet=1 [accessed December 27, 2006].

Spotswood S. 2005. HHS flu plan aims to lift vaccine supply. *U.S. Medicine*. [Online]. Available: http://www.usmedicine.com/article.cfm?articleID=1210&issueID=82 [accessed January 26, 2007].

St. John RK, King RA, de Jong D, Bodie-Collins M, Squires SG, Tam TW. 2005. Border screening for SARS. *Emerging Infectious Diseases* 11(1):6-10.

Stern AM, Markel H. 2004. International efforts to control infectious diseases. 1851 to the present. *Journal of the American Medical Association* 292(12):1474-1479.

Stier DD, Goodman RA. In Press. Mutual aid agreements—essential tools for public health preparedness and response. *American Journal of Public Health*.

Stohr K, Esveld M. 2004. Will vaccines be available for the next influenza pandemic? *Science* 306(5705):2195-2196.

Sutton V. 2001 (Summer). Bioterrorism preparation and response legislation—the struggle to protect states' sovereignty while preserving national security. *The Georgetown Public Policy Review* 93.

Taubenberger JK, Reid AH, Lourens RM, Wang R, Jin G, Fanning TG. 2005. Characterization of the 1918 influenza virus polymerase genes. *Nature* 437(7060):889-893.

Thomas DB, Arita I, McCormack WM, Khan MM, Islam S, Mack TM. 1971. Endemic smallpox in rural East Pakistan. II. Intravillage transmission and infectiousness. *American Journal of Epidemiology* 93(5):373-383.

Thomson AK, Faith K, Gibson JL, Upshur RE. 2006. Pandemic influenza preparedness: An ethical framework to guide decision-making. *BCM Medical Ethics* 7(1):E12.

Thorson AM, Petzold M, Nguyen TKC, Ekdahl K. 2006. Is exposure to sick or dead poultry associated with flulike illness? A population-based study from a rural area in Vietnam with outbreaks of highly pathogenic avian influenza. *Archives of Internal Medicine* 166(1):119-123.

Torda A. 2006. Ethical issues in pandemic planning. *Medical Journal of Australia* 185(10): S73-S76.

Tumpey TM, Basler CF, Aquilar PV, Zeng H, Solorzano A, Swayne DE, Cox NJ, Katz JM, Taubenberger JK, Palese P, Garcia-Sastre A. 2005. Characterization of the reconstructed 1918 Spanish influenza pandemic virus. *Science* 310(5745):77-80.

UN (United Nations). 1985. Annex. In International Covenant on Civil and Political Rights (E/CN.4/1985/4).

University of Toronto Joint Centre for Bioethics Pandemic Influenza Working Group. 2005. *Stand on Guard for Thee: Ethical Considerations in Preparedness Planning for Pandemic Influenza*. Toronto, Canada: University of Toronto Joint Centre for Bioethics. [Online]. Available: www.utoronto.ca/jcb/home/documents/pandemic.pdf [accessed December 27, 2006].

United Nations Economic and Social Council. 1985. *Siracusa Principles on the Limitation and Derogation Provisions in the International Covenant on Civil and Political Rights*.

Upshur RE. 2002. Principles for the justification of public health intervention. *Canadian Journal of Public Health* 93(2):101-103. [Online]. Available: http://www.fhs.mcmaster.ca/ceb/community_medicine_page/docs/upshur.pdf [accessed December 27, 2006].

Upshur RE. 2003. The ethics of quarantine. *American Medical Association Journal of Ethics Virtual Mentor* 5(11). [Online]. Available: www.ama-assn.org/ama/pub/category/11535.html [accessed December 27, 2006].

U.S. Congress. 1946. The Public Health Service Act. USC Title 42, Chapter 6A, Part G.

U.S. Congressional Budget Office. 2006. *A Potential Influenza Pandemic: An Update on Possible Macroeconomic Effects and Policy Issues*. [Online]. Available: http://www.cbo.gov/ftpdocs/72xx/doc7214/05-22-Avian%20Flu.pdf [accessed January 26, 2007].

U.S. Census Bureau. 2000. *Census 2000*. [Online]. Age distribution was retrieved from www.census.gov/census2000/states/us.html, and household sizes were retrieved from http://factfinder.census.gov/servlet/DatasetMainPageServlet?_lang=en (see Census 2000 Summary File 1). [accessed October 29, 2002].

Watts DJ. 1999. *Small Worlds: The Dynamics of Networks Between Order and Randomness*. Princeton, NJ: Princeton University Press.

Watts DJ, Strogatz S. 1998. Collective dynamics of "small-world" networks. *Nature* 393(6684):440-442.

White House. 2005 (April 1). *Executive Order: Amendment to E.O. 13295 Relating to Certain Influenza Viruses and Quarantinable Communicable Diseases*. [Online]. Available: http://www.whitehouse.gov/news/releases/2005/04/20050401-6.html [accessed December 27, 2006].

White House Homeland Security Council. 2005. *National Strategy for Pandemic Influenza*. [Online]. Available: http://www.whitehouse.gov/homeland/nspi.pdf [accessed January 26, 2007].

Whittle P. 1955. The outcome of a stochastic epidemic. *Biometrika* 42:116-122.

WHO (World Health Organization). 2003. *Hospital Infection Control Guidance for Severe Acute Respiratory Syndrome (SARS)*. [Online]. Available: http://www.who.int/csr/sars/infectioncontrol/en/ [accessed January 26, 2007].

WHO. 2004. *WHO SARS Risk Assessment and Preparedness Framework*. [Online]. Available: http://www.who.int/csr/resources/publications/CDS_CSR_ARO_2004_2.pdf [accessed January 26, 2007].

WHO. 2005a. *WHO Checklist for Influenza Preparedness Planning*. Sec 1.5.2. Geneva: World Health Organization. [Online]. Available: www.who.int/csr/resources/publications/influenza/FluCheck-6web.pdf [accessed December 27, 2006].

WHO. 2005b. Avian influenza A (H5N1) infection in humans. The Writing Committee of the World Health Organization Consultation on Human Influenza A/H5. *New England Journal of Medicine* 353(13):1374-1385.

WHO. 2005c. *Avian Flu vs. Pandemic Flu*. [Online]. Available: http://www.wvdhhr.org/healthprep/common/avian_vs_pandemic_flu.pdf [accessed January 26, 2007].

WHO. 2005d. *Avian Influenza: Assessing the Pandemic Threat*. [Online]. Available: http://www.who.int/csr/disease/influenza/H5N1-9reduit.pdf [accessed January 26, 2007].

WHO. 2006a. *Avian Influenza—Spread of the Virus to New Countries*. [Online]. Available: http://www.who.int/csr/don/2006_02_21b/en/index.html [accessed January 26, 2007].

WHO. 2006b. *Avian Influenza: Significance of Mutations in the H5N1 Virus*. [Online]. Available: http://www.who.int/csr/2006_02_20/en/index.html [accessed January 26, 2007].

WHO. 2007a. *Situation Updates—Avian Influenza*. [Online]. Available: http://www.who.int/csr/disease/avian_influenza/updates/en/index.html [accessed January 26, 2007].

WHO. 2007b. *Cumulative Number of Confirmed Human Cases of Avian Influenza A/(H5N1) Reported to WHO*. [Online]. Available: http://www.who.int/csr/disease/avian_influenza/country/cases_table_2007_05_16/en/index.html [accessed May 16, 2007].

Wong SS, Yuen KY. 2006. Avian influenza virus infections in humans. *Chest* 129(1):156-168.

World Health Assembly. 2005. *Strengthening Pandemic Influenza Preparedness and Response*. WHA 58:13.9. [Online]. Available: http://www.who.int/csr/disease/influenza/A58_13-en.pdf [accessed January 26, 2007].

World Health Organization Writing Group. 2006. Nonpharmaceutical interventions for pandemic influenza, international measures. *Emerging Infectious Diseases* 12(1):81-87. [Online]. Available: http://www.cdc.gov/ncidod/EID/vol12no01/05-1370.htm [accessed January 26, 2007].

ANNEX 3-1 follows

TABLE 3-4 Daily Transmission Probabilities, x,[a] Among Children and Adults, by Mixing Group, and Group Sizes

Contact group		Mean size	Children: Pre-school: Small playgroup	Children: Pre-school: Large daycare	Children: School: Elementary	Children: School: Middle	Children: School: High	Adults
Small playgroupd		2.9	0.03000					
Large day-care centers		15.8		0.02000				
Elementary school		77.8			0.01000			
Middle school		145.3				0.00800	0.00800	
High school		113.7						
Family		2.5						
	Child		0.03520	0.03520	0.03520	0.03520	0.03520	0.01240
	Adult		0.01240	0.01240	0.01240	0.01240	0.01240	0.01510
Household social cluster		10.1						
	Child		0.03000	0.03000	0.03000	0.03000	0.03000	0.01000
	Adult		0.01000	0.01000	0.01000	0.01000	0.01000	0.01000
Hospital								
	Smallpox ward	133.0						
	Worker-worker							0.00200
	Worker-visitor		0.00200	0.00200	0.00200	0.00200	0.00200	0.00200
	Patient-worker		0.00010	0.00010	0.00010	0.00010	0.00010	0.00010
	Patient-visitor		0.00010	0.00010	0.00010	0.00010	0.00010	0.00010
	Other wards	533.0						0.00050
Workgroup								0.01000
Neighborhood		500.0	0.00004	0.00004	0.00005	0.00005	0.00005	0.00014
Community		2000.0	0.00001	0.00001	0.00001	0.00001	0.00001	0.00003

[a]The probability that an infected person with ordinary smallpox, on the second day after the onset of fever, makes sufficient contact to infect an unvaccinated susceptible person in the mixing group being modeled.

TABLE 3-11 Summary of Results of Epidemic Simulation Runs Showing the Effects of "No Response" Scenarios 1 and 2 and Response Scenarios 3-10 on Epidemics Initiated by the Introduction of Ten Smallpox Cases into 6,000-Person Towns

	Single Uniform Town				Four Hub-and-Spoke Towns				Ring of Six Towns			
	Cases	Deaths	New Vaccina-tions	Duration (days)	Cases	Deaths	New Vaccina-tions	Duration (days)	Cases	Deaths	New Vaccina-tions	Duration (days)
Scenario	Mean (SD)	Mean (SD)	Mean (SD)	Mean (SD)	Mean (SD)	Mean (SD)	Mean (SD)	Mean (SD)	Mean (SD)	Mean (SD)	Mean (SD)	Mean (SD)
No response												
1	5,231.6	1,650.4	NA	649.7	5,226.9	1,638.9	NA	643.1	5,240.2	1638.7	NA	629.1
	(46.5)	(34.7)	(NA)	(124.4)	(58.3)	(40.0)	(NA)	(146.1)	(47.5)	(30.6)	(NA)	(115.0)
2	2,981.3	935.0	NA	394.9	2,521.2	788.1	NA	470.9	2,167.9	676.4	NA	416.2
	(623.2)	(198.3)	(NA)	(82.2)	(583.0)	(181.5)	(NA)	(125.5)	(996.3)	(313.0)	(NA)	(125.3)
Response												
3	47.2	11.3	355.7	76.9	35.5	9.5	295.0	69.2	43.6	11.4	332.9	75.6
	(16.5)	(5.1)	(88.2)	(17.7)	(14.4)	(4.9)	(78.4)	(15.1)	(14.8)	(4.9)	(68.8)	(16.5)
4	45.9	11.4	351.6	73.0	39.9	9.3	309.7	70.5	34.8	8.9	306.8	68.5
	(21.9)	(5.8)	(99.2)	(17.7)	(16.6)	(4.2)	(69.5)	(16.5)	(12.4)	(4.2)	(64.6)	(16.1)
5	35.9	9.8	316.2	66.9	41.2	11.0	324.6	71.4	38.5	10.3	314.9	65.6
	(12.7)	(4.2)	(62.0)	(12.6)	(11.8)	(3.7)	(60.2)	(13.2)	(17.7)	(5.0)	(75.9)	(10.4)
6	32.2	8.4	2,373.1	59.7	28.9	7.0	2,359.8	57.9	24.2	6.8	2,357.9	54.8
	(13.6)	(4.5)	(46.1)	(12.7)	(10.4)	(3.1)	(45.3)	(11.1)	(6.7)	(3.0)	(44.5)	(10.6)
7	25.1	7.0	2,364.6	58.3	29.4	7.5	2,373.9	60.0	28.2	7.8	2,363.0	58.3
	(8.6)	(3.5)	(49.7)	(13.8)	(8.9)	(2.6)	(42.4	(9.9)	(10.70	(3.4)	(51.9)	(15.9)
8	22.2	5.8	4,502.7	46.9	19.8	4.7	4,492.5	46.7	17.5	4.9	4,505.0	43.3
	(10.1)	(2.9)	(28.9)	(6.4)	(7.2)	(2.6)	(31.3)	(10.2)	(4.0)	(1.9)	(29.5)	(7.4)
9	17.3	4.3	4,505.7	45.2	19.9	5.9	4,501.8	46.7	18.6	5.1	4,504.1	45.2
	(4.7)	(2.0)	(38.4)	(8.0)	(5.6)	(2.1)	(38.0)	(8.1)	(5.8)	(2.6)	(29.80)	(5.8)
10	44.2	11.2	334.9	70.2	42.5	10.9	323.5	75.5	32.5	8.2	282.9	66.1
	(21.1)	(5.5)	(94.0)	(18.3)	(17.9)	(5.3)	(82.4)	(19.5)	(10.2)	(3.9)	(46.5)	(10.0)

Data from single uniform town, ring town, and hub-and-spoke town architectures are shown. For each scenario, the number of infected persons, number of deaths, number of vaccinations administered, and duration of the epidemic are shown. Totals include index generation (G_0) cases along with subsequent cases.

TABLE 3-12 Summary of Results of Epidemic Simulation Runs Showing the Effects of "No Response" Scenarios 1 and 2 and Response Scenarios 3–10 on Epidemics Initiated by the Introduction of 500 Smallpox Cases into 50,000-Person Towns

	Single Uniform Town				Ring of Six Towns			
	Cases	Deaths	New Vaccinations	Duration (days)	Cases	Deaths	New Vaccinations	Duration (days)
Scenario	Mean (SD)	Mean (SD)	Mean (SD)	Mean (SD)	Mean (SD)	Mean (SD)	Mean (SD)	Mean (SD)
No response								
1	46,474.6	14,641.7	NA	1,280.8	46,463.7	14,623.4	NA	1,204.5
	(109.3)	(117.9)	(NA)	(159.6)	(100.0)	(66.2)	(NA)	(139.3)
2	26,915.2	8,435.3	NA	453.2	27,103.0	8,476.0	NA	531.7
	(819.4)	(288.6)	(NA)	(30.4)	(784.3)	(220.60	(NA)	(77.2)
Response								
3	1,609.5	415.6	8,780.7	104.3	1,563.6	406.8	8,325.4	105.1
	(90.2)	(29.6)	(445.2)	(13.1)	(95.4)	(28.9)	(407.7)	(16.6)
4	1,635.9	429.9	8,945.1	109.2	1,500.2	400.5	8,114.5	107.3
	(133.5)	(35.0)	(592.5)	(18.6)	(59.2)	(16.1)	(445.6)	(10.1)
5	1,657.3	429.6	8,988.3	112.5	1,472.6	392.1	7,993.1	101.9
	(88.0)	(22.5)	(371.2)	(9.7)	(126.7)	(37.1)	(618.2)	(11.8)
6	1,146.9	305.1	22,023.1	87.1	1,101.2	299.4	21,723.2	81.4
	(69.3)	(21.1)	(149.3)	(6.5)	(55.8)	(13.7)	(202.6)	(6.3)
7	1,160.2	310.6	22,041.2	86.8	1,088.2	295.1	21,784.2	90.5
	(52.5)	(20.6)	(206.0)	(6.2)	(43.2)	(15.3)	(202.6)	(11.5)
8	772.5	216.4	37,516.9	66.0	775.0	219.9	37,529.3	65.6
	(45.7)	(14.0)	(84.8)	(8.9)	(17.4)	(11.6)	(115.3)	(10.3)
9	780.5	211.7	37,560.0	63.6	756.8	214.8	37,569.0	60.3
	(25.6)	(12.1)	(59.6)	(7.4)	(32.1)	(10.5)	(117.8)	(5.0)
10	1,634.8	428.2	8,928.2	116.0	1,512.6	398.3	8,018.1	104.5
	(79.3)	(21.6)	(336.2)	(14.9)	(50.1)	(20.2)	(248.3)	(11.1)

Data from simulations on the single uniform town and ring town architectures are shown (simulations on the hub-and-spoke town architecture were not performed for the 50,000-person town).

4

Ethical Issues in Pandemic Planning and Response

OVERVIEW

Many of the conferences, meetings, and workshops convened in anticipation of an H5N1 influenza pandemic have focused on the specific strategies that can be used in fighting such a pandemic. The contributors to this chapter take a different tack and consider the creation of ethical guidelines for governments, health-care systems, and clinicians to be used in planning for and responding to a pandemic. The authors identify a set of ethical principles that should serve as a foundation for such guidelines; they also discuss the importance of public engagement in the development of the guidelines and the need for clear communication of the guidelines once they are done.

In the first contribution to this chapter, Alexander Capron of the University of Southern California examines a variety of ethical approaches to pandemic planning, noting that ethics may be applied to both the content of policies and the processes by which they are established and implemented. Returning to the central ethical considerations identified by his former World Health Organization (WHO) colleague, David Heymann (see page 33), Capron specifically addresses the implications of pandemic influenza for human rights, access to health care, obligations of and to health-care workers, and obligations of countries and intergovernmental organizations. He then explores how ethical principles can be applied in policy making to address these issues. Capron's observation that all dilemmas faced by pandemic planners can be reduced to "the classic struggle between individual and group" echoes Victoria Sutton's conception of a "pandemic flu ethic," which she defined in her remarks at the workshop as "a limitation on the freedom of action or the imposition of a duty to act in the pursuit of the continued existence of life and order." Sutton, who has taken a similar approach to defining an ethic of biodefense and bioterrorism (Sutton, 2005) views ethics as a precursor to law (see also Chapter 3 and the Summary and Assessment), while Capron portrays ethics as integral to public health policy.

Capron's essay also features "straightforward, practical suggestions" for pandemic preparations that are supported by ethical principles. At the national level he advocates advanced planning, communication, and public involvement in order to realize an "ethically responsible and appropriate response." At the international level, he calls on governments of wealthier nations to announce support for poor, early-affected countries out of both ethical responsibility and self-interest. Even if public debate results in differences in pandemic policies among communities and countries, Capron contends, civic engagement will promote the understanding and acceptance of necessarily imperfect—but beneficial—public health measures.

Focusing on the disproportionate burden that a pandemic is likely to place on the world's poorest people and countries, Ruth Faden of Johns Hopkins University asserts in her contribution to this chapter that "the greatest moral challenge posed by a pandemic is how to respect commitments to social justice in the face of the overwhelming and entrenched inequalities." Such inequalities result from efforts to control avian influenza that disproportionately burden poor countries and benefit wealthy ones, and they are also likely to result from an exacerbation of social injustice within the U.S. and other wealthy countries in the response to a pandemic. Therefore, Faden argues, governments bear a moral responsibility to identify where social injustices are likely to occur as the result of a pandemic and to take reasonable steps to prevent or reduce the worst among them. In order to support this effort, Faden and fellow members of the Bellagio Group have developed a set of principles that are intended to uphold the rights and interests of disadvantaged groups in pandemic planning and response as well as a set of checklists to guide the incorporation of these principles into pandemic planning and response. In her essay, Faden describes these principles, the rationale behind them, and their significance to public health policy and practice.

As noted in the Summary and Assessment, several workshop participants raised concerns regarding the lack of clear authority for decision-making in public health emergencies. Sutton has described the history and consequences of the longstanding conflict between federal and state claims to public health authority and has suggested a potential resolution through a system of "cooperative federalism" in which the federal government establishes standards for pandemic measures that are subsequently administered by state government and implemented at the local level (Sutton, 2001). This model of federal leadership is endorsed and expanded upon in the chapter's third essay, by speaker Shelley Hearne of Johns Hopkins University, who argues that "from an ethical standpoint, federal health agencies should play a more directive role in establishing standards and critical requirements for state and local jurisdictions in order to ensure equal levels of preparedness for all citizens." She presents a detailed strategy for building emergency-response capacity that will protect all U.S. citizens equally and, in much the same way that the federal government ensures that each state meets certain minimal environmental standards, will enforce basic requirements for public health in pandemic preparations and response.

The chapter concludes by shifting from the big picture of government action in a pandemic to the smaller scale of the clinic, where the personal agony of ethical dilemmas comes into focus. Speaker Bernard Lo of the University of California, San Francisco, and co-author Douglas White confront the need for specific criteria to triage patients with respiratory failure in the likely event of a shortage of respirators during a pandemic as well as the need for guidelines and procedures to address the practical problems that will arise when such policies are implemented. The authors stress that public participation in the crafting of such guidelines and procedures will be important in creating popular acceptance of the difficult choices that must be made during a pandemic.

While recognizing the ideal of public participation in pandemic planning, workshop participants agreed that public health professionals must expect that most people will be entirely unprepared when the next pandemic strikes. In order to mount an effective response, public health authorities will need to act rapidly and authoritatively on the basis of incomplete knowledge. To the biomedical experts who would inform these decisions, Institute of Medicine (IOM) president Harvey Fineberg posed a series of rhetorical questions: Are experts bound to frame evidence, based on their knowledge, so that politicians reach "correct" conclusions regarding a threatened pandemic? Should experts refrain from making conclusions, but merely answer questions? Should experts speak directly to the media about their concerns? Rather than offering answers, Fineberg described how various experts approached these dilemmas in the course of reacting to the appearance of swine flu in 1976 and how those reactions—and their treatment in the media—shaped the nation's response to a threatened pandemic.

One television network, NBC, provided coverage that was sympathetic to the federal program of mass vaccination against swine flu, while another network, CBS, offered skepticism and criticism of the government's actions, Fineberg recalled. Such a contrast in interpretation was rare in the media at that time, he said, and it originated in the distinct pool of experts that each network consulted on the story. When White House contacts told NBC reporters that the program was being carried out despite its disadvantage to President Ford, who was up for reelection, the network concluded that the President was being forced to do the bidding of scientific experts. At the same time, CBS reporters heard from their Centers for Disease Control and Prevention (CDC) contacts that the vaccination program was premature and unfeasible, leading that network to conclude that the vaccination program was being launched solely for political reasons. Given the potential to create similarly influential and divisive messages in the face of pandemic influenza, experts should think carefully about their roles and responsibilities in portraying their convictions, understandings, and beliefs to the media, Fineberg advised.

But no matter what choices are made to address a threatened pandemic, there will be skepticism, criticism, and differences of opinion, Fineberg concluded. Because "there is no way to avoid the dilemmas posed by acting without full scientific knowledge," as Gostin has observed, "the only safeguard is the adoption of ethical values in formulating and implementing public health decisions" (Gostin, 2004).

ETHICAL CONSIDERATIONS IN INTERNATIONAL PREPAREDNESS PLANNING EFFORTS

Alexander Morgan Capron
University of Southern California[1]

Earlier papers have detailed the public health history of past epidemics, from polio to SARS, and have described how health-care professionals, particularly in public health, are organized to respond to existing and emerging communicable diseases. With this background we can now move to a discussion of the ethical considerations in preparedness planning efforts.

The first question we encounter when thinking about these ethical considerations is, where exactly does ethics fit into international preparedness planning efforts? Not surprisingly, my view is that it must lie at the heart of the process because it helps us see what the right thing is to do under a particular set of circumstances. But in doing so ethical analysis must examine both the substance and the consequences of alternative policies and practices and the processes by which they are developed and selected. In this essay I will introduce some of the ethical considerations relevant to pandemic influenza planning but will not attempt to cover them all. In particular, I will leave some of the ethical issues raised by disease mitigation and resource allocation to be addressed when those topics are specifically discussed. But before talking about how ethics can be applied to pandemic planning, it seems advisable to say a few words about what "ethics" consists of.

[1]These remarks grow out of work on which I was engaged (until August 2006) as Director, Department of Ethics, Trade, Human Rights and Health Law, at WHO. These efforts are being carried forward in the department by Dr. Andreas Reis, in collaboration with our colleagues in the Department of Epidemic and Pandemic Alert and Response, especially the pandemic influenza team leader, Dr. Keiji Fukuda, and department director, Dr. Mike Ryan, as well as the Acting Assistant Director-General for the Communicable Diseases Cluster, Dr. David Heymann. In these remarks, however, I write only for myself and not for WHO.

The Range of Ethical Theories

Ethical Theories Relevant to Policies and Practice

1. Deontology and principilism. How does ethics provide a guide to right action? Many ethical theories are used in medicine and public health, including deontology (in which decisions are based mainly on a consideration of one's duties), casuistry, consequentialism, virtue ethics, and rights-based theories. Of these, the first is probably the most familiar to people working in health care—and has been since the time of Hippocrates. His oft-repeated oath and his many other injunctions, such as "First, do no harm" (familiar to generations of physicians in its Latin version, "*Primum non nocere*"), form the basis for a set of professional obligations which consists primarily of doctors' duties toward their patients.

Another, more contemporary statement of the principles that should guide health professionals was set forth in the Belmont Report, produced in 1978 as the capstone of the work of the National Commission for the Protection of Human Subjects of Biomedical and Behavioral Research. Afterwards, the principles were elaborated on by Tom Beauchamp, who had been the principal consultant on the Belmont Report, and his colleague Jim Childress in *Principles of Biomedical Ethics*, now in its fifth edition (Beauchamp and Childress, 2001).

The two ethicists set fourth four principles—beneficence, non-maleficence, respect for persons, and justice—that have been used so widely in discussions regarding health-related topics that they may sometimes be mistaken as all that bioethics has to say on the subject. Indeed, they are invoked so formulaically in so many settings that they are sometimes mocked as "the Georgetown mantra" (referring to Georgetown University, at whose Kennedy Institute of Ethics both Beauchamp and Childress were teaching when they wrote their book). These four principles are indeed often useful, but they ought not to be substituted for careful ethical analysis, both because the principle most relevant to public health actions—namely justice—is the least fully discussed in the literature applying the principles to medical practice and biomedical research and also because the list of the four principles itself does not provide guidance on how to give more weight to one than the other when they are in conflict.

The Belmont Report and its kin are a form of principilism, or making ethical decisions based on a set of principles. Principilism is one way of approaching professional deontology, particularly in the case of health-care clinicians and researchers. Although experience indicates that these principles are indeed useful as a guide for individual physicians and researchers in thinking about their obligations to individual patients and subjects and in understanding the rules that appear in many codes of ethics or in determining when further rules are needed, we should keep in mind that this is only one set of principles among many.

Rights-based ethics, for example, is an alternative formulation that involves a larger number of principles and is addressed more to the actions of institutions and governments. An example of rights-based ethics is the *Universal Declaration on Bioethics and Human Rights,* adopted by the General Conference of the United Nations Educational, Scientific and Cultural Organization (UNESCO) in October 2005. As suggested by its name, rights-based ethics attempts to ground its principles not in philosophic or medical tradition but in human rights, particularly those that have been agreed upon in international conventions. People in the human-rights field point out that these are not vague ethical obligations but rather are binding legal rights. Like all matters that arise out of intergovernmental debate and consensus, the principles in UNESCO's bioethics declarations are written with a great deal of imprecision (in order to get countries with markedly different views to agree to them), and, like the Belmont Report, they also are not particularly helpful in those inevitable situations where principles collide or, sometimes, even point in opposite directions.

2. Consequentialism and "the greatest good." Of the alternatives to principilism, the ethical approach of greatest relevance for public health actions is probably consequentialism, the most influential example of which is utilitarianism. Whereas a deontologist says that people should act so as to fulfill their duties to others, a consequentialist says that a person should act in the way that produces the best outcome. Of course, what counts as the best outcome varies among ethical viewpoints. For a utilitarian, at least a utilitarian of a Benthamite stripe, the right action is that which produces the greatest sum of pleasure in the relevant population, or more generally that which maximizes human welfare or well-being. A related approach, "rule utilitarianism," looks at the consequences of general rules instead of the consequences of individual acts. For a rule utilitarian, the question is not whether a particular act will produce the greatest good but instead whether a particular rule—say, "Do not lie"—as a general matter will produce the greatest good even if there may be situations in which not following the rule (e.g., by lying) would produce a better outcome. By reasoning in this way, obligations can be generated on consequentialist grounds.

Even when people agree on the "facts" about a set of policies or practices, they may reach differing conclusions about whether the policies are ethically justifiable, and often the reason for the different conclusions is that the people are applying different ethical approaches. Consider, for example, a decision to allocate limited supplies of antiviral medicines to hospitalized patients who were already very sick before they contracted influenza instead of providing the medicines to people in their homes who were found by community nurses to be showing early signs of influenza. Such a decision might seem appropriate to a deontologist who believes that the duty of beneficence attaches to all existing physician-patient relationships and who therefore reasons that it is the physician's duty to help the infected patients fight the influenza. On the other hand, the decision might seem wrong to a utilitarian who calculated that many more lives would be saved by treating otherwise healthy outpatients who were just developing influenza rather than inpatients with co-morbidities who might not survive even if the influenza treatments were successful. The difference of opinion arises not from any disagreement over what the outcomes of the two choices would be but rather from a disagreement about which factors are important in judging ethical actions or about the weights that should be accorded to the various factors. A principilist, for example, might ask which outcome is more important, achieving justice or respecting persons, while a utilitarian might want to know which changes would do the most to increase total welfare. Sometimes it is possible to apply more than one ethical system to a set of policies and thereby meet the concerns of people with different views of what is right—for example, to select among policies that have all passed muster on utilitarian grounds, the one policy that is fairest or that best promotes self-determination by patients.

Ethical Standards for the Process

Besides evaluating the content of pandemic preparations and response plans, one can also examine the process by which such policies are established. Again, such an evaluation can rely on duty-oriented considerations, outcomes-based considerations, or both. It is now widely agreed, both on philosophical grounds (e.g., what Prof. Norman Daniels calls "fair process") and on human-rights grounds (e.g., the obligations of governments to resident populations), that the process by which these policies are decided should conform to certain standards. Perhaps the most important is that the people who will be affected by the policies should be kept informed and be allowed to participate in deliberations about the policies through processes in which the reasons, principles, and evidence that they regard as relevant are considered. Furthermore, the decision-making process should allow for revisiting and revising policies in light of new evidence and arguments as well as for formal appeals of the policies. Finally, there should be mechanisms to ensure that these criteria are actually fulfilled.

The Values Embedded in Policies

In their analyses of the moral reasons for formulating a policy or taking an action, ethicists are not limited to looking at the ethical principles that were included explicitly in the process; they can also point out values that were included implicitly in what might otherwise seem to be purely technical decisions. At WHO, for example, the processes for developing and promulgating standards typically involve consultations with experts as well as approval by WHO's governing bodies, but until recently the processes have not also overtly included any ethical analyses. Nonetheless, it has always been the case that whenever WHO developed guidance in any of the many areas where it sets technical norms, it was also implicitly importing value preferences and adopting ethical norms.

The same is true for WHO's member states and for other major actors, such as health-related nongovernmental organizations. Accordingly, one of the things we have done at WHO is to look at the pandemic preparedness plans that are being produced by countries around the world to see what value assumptions they reveal. Few of these plans identify the particular ethical considerations used in their formulations, so we have had to try to unpack the values revealed by the specific policy choices that were made.

Consider, for instance, a pandemic program plan that gives its goal as saving the most lives. This might seem so straightforward as to be beyond ethical dispute—after all, it is a goal shared with most medical institutions and public health authorities. Yet in the context of a particular health system, such an objective might translate into a preference for treating particular groups of patients.

To illustrate this, consider an example from another arena, namely the choices facing countries participating in the so-called 3-by-5 Initiative of the Joint United Nations Programme on AIDS (UNAIDS) and WHO: If treatment programs had targeted the patients who were the easiest to treat, it would have been possible to make much more progress toward the initiative's aim of having three million AIDS patients in developing countries on antiretroviral (ARV) treatment by the end of 2005. But targeting those easiest to reach might result in focusing the initiative's resources on those who were already privileged in various ways: those who have good access to health care (because they live near a major medical center capable of operating ARV treatment programs and they have sufficient funds to obtain care there); those who are best informed about HIV/AIDS in general and about how to obtain treatment; those who are otherwise healthiest and hence find it easiest to seek care; or those who are best protected against the negative consequences of a positive HIV diagnosis (such as loss of job, family rejection, etc.) and hence most likely to have sought out and received an HIV test and therefore have discovered that they needed treatment.

If utility—measured simply in terms of lives saved—is the principal or sole measure of how ethical a policy is, then using limited resources to successfully treat the largest number of patients possible would seem justified even if that meant preferring those with other advantages. After all, treating such patients would likely cost the least per life saved and therefore make it possible to save the most people. But there are other good reasons—turning on fairness rather than utility—why a government ought not simply devote its treatment resources to those who already had the best care, who were already the healthiest, who were already close to hospitals with ARV clinics, who had the means to partially pay for their treatments, or who knew the most about their need for treatment. In particular, focusing solely on utility would seriously disadvantage poor, rural, female, and socially marginalized populations, and considerations of justice would imply that at least some of the government's treatment budget should be spent on reaching out to remote or socially isolated populations by providing community education, creating clinics, or perhaps even building basic infrastructure, such as roads to remote villages. The result would be to modify the goal of saving the most lives with the "side constraint" of treating all people with equal need in an equal fashion. No matter what balance of objectives is ultimately chosen, a process that explicitly recognizes these considerations is likely to be more justifiable than one that pretends that value choices are not being made and that produces a plan based only on "technical" considerations.

In both our work on HIV (see *Guidance on Ethics and Equitable Access to HIV Treatment and Care* [UNAIDS/WHO, 2004]) and our work with pandemic influenza, our ethics team has insisted on working in tandem with the departments that are involved in the technical work, which, in the latter case, is principally the Department of Epidemic and Pandemic Alert and Response. The idea is that ethical guidance and analysis should not stand on its own, but rather it should be incorporated into and shape the processes by which the technical norms and standards for treatment, prophylaxis, and other influenza-related public health measures are developed, both by WHO and by WHO's member states. In this way ethical guidance is explicitly and deeply embedded in advice that might otherwise be regarded as being merely technical.

Ethics and Prudence

While we may speak of our duties to ourselves ("You owe yourself a break—you've been working too hard!"), and while choices that affect us can also be framed in terms of our obligations to other people ("You shouldn't take risks like that or you'll leave your children as orphans") or to the deity ("It would be wrong to take your own life, which is a gift from God"), most ethical duties are framed in terms of what effects our behavior can have for others. Sometimes, however, doing something to benefit others is framed not as something we should do because we have agreed to do so (for example, in a human rights convention) or because doing so is inherent in our role (for example, a Hippocratic duty) or even because doing so would maximize overall human welfare, but rather because it would be the smart thing to do to achieve benefits for ourselves. Of course, a single act can serve multiple aims—the Marshall Plan, for instance, was a great humanitarian effort that responded to the desperate conditions in a ravaged continent after the Second World War, but it also served to build a strategic buffer against the expansion of Soviet power. Such a mixture of aims can complicate analysis, but it is always important to distinguish an argument that we should do something because it will be to our material benefit (which is a prudential claim) from arguments based on something being the morally necessary or morally desirable thing to do (which is an ethical claim).

Four Realms of Influenza Preparedness Planning and Action

At WHO the ethical issues we have been dealing with concerning influenza preparedness planning and action can be divided into four groups or areas of concern. These four areas were also apparent in David Heymann's review of the history of public health responses to communicable disease outbreaks in Chapter 1, and, indeed, the history that Heymann describes provides a great deal of material with which to evaluate the ethical arguments for or against certain policies or actions. So let us begin by reviewing those four areas and then investigate how ethical considerations arise in each.

Four Areas of Ethical Concern

The first area of concern is equitable access to health care in a pandemic. A major issue here is how influenza vaccines, antivirals (Tamiflu), and hospital beds should be allocated for influenza patients. More broadly, the question is how health-care resources in general should be allocated, both before and during the pandemic, between the needs of the influenza effort and other health needs. A related, more specific question is whether it is appropriate to alter the standards for approval of vaccines or drugs for a pandemic because of the pressing public need.

The second area of concern is the ethics of public health actions taken in response to a pandemic, such as the surveillance of outbreaks of animal and human pathogens and dissemination of outbreak information; measures to prevent animal-to-human transmission through culling of livestock and so forth; separation measures such as quarantine, isolation, and social distancing; and control of international travel and borders, partly in response to the new WHO International Health Regulations.

The third area of ethical concern is the obligations of health-care workers during a pandemic and the obligations of society to them in return. As Dr. Heymann made clear, a notable feature of many of the outbreaks he reviewed was the danger to and mortality among the health-care professionals dealing with those outbreaks, including both those who were providing treatment and those who were simply monitoring the outbreak. One can assume that if health-care workers are at greater-than-ordinary risk for acquiring infections because of their jobs—which appeared to be the case with SARS and may or may not be the case with pandemic influenza—then their natural inclination to minimize their exposure would be in conflict with their professional obligations both to individual patients and perhaps to their communities as a whole. Their acceptance of this risk in the execution of their duties would engender reciprocal duties on the part of the community to them.

When we speak of health-care professionals' duties—perhaps most fully articulated in the case of physicians, but certainly recognized for other professionals as well—we need to ask whether such duties derive from their special training and their status as licensed, self-governing professions, or whether they reflect the fact that they possess a set of skills that are particularly needed under the circumstances. And if it is the latter, would this rationale not also extend beyond workers who fit the usual, narrow definition of professionals? If "possessing essential skills" is the criterion, then would this class not encompass other people who are central to the operation of the health-care system, especially in times of crisis, right down to the delivery drivers, the maintenance workers who keep facilities going and clean up patient rooms, and so forth?

A further question here is whether the obligations of health-care workers are dependent upon their receiving any special protection from society. If the obligations are to be linked to special protection, it would imply a contractual/reciprocal model of their roles rather than a professional model, in which certain duties are inherently part of the job.

Since various ethical issues concerning the role of health-care workers are disputed—both because of questions about the relevance of ancient precepts to modern practitioners and because of the many workers who are not professionals in the same sense as physicians—the social-contract model may prove useful in deciding whether to give these groups special status when allocating prophylaxis and treatment for pandemic influenza and, if so, why. Do these workers have any special claim? One might say no on the grounds that, just as soldiers who sign up in peacetime are obligated to serve during war, workers in health care go into the field knowing that it has some unavoidable risks. The justice of concluding that they should not get special treatment on the occasions when these risks actually arise is reinforced—at least as to physicians and, to a lesser extent, nurses—by their having received a very heavily subsidized education that put them in a very privileged position in society.

The fourth area of ethical concern centers on obligations among countries and the obligations of intergovernmental organizations: How should governments balance their duties to their own populations versus duties to other countries and populations, and what role should international organizations such as WHO play in addressing the cross-border risks and obligations?

The first of these questions is closely related to an issue raised in a workshop discussion of vaccine-allocation strategies (see Summary and Assessment): How should decision makers determine the appropriate point in time to release a particular portion of the preventive and curative medicines under their control? Here the question is whether it is appropriate for a country to release scarce supplies of vaccine or other treatment to a second country when that second country is experiencing a pandemic, instead of holding on to the supplies for possible use by its own population. If the answer is yes, then one must also ask at what point in time should the release be made. If the first country holds back and the epidemic is contained—particularly if this is due to the aid of still other countries—then the decision makers will be seen as not having responded as dictated by humanitarian principles and perhaps human rights obligations and thus being responsible for a loss of lives that could have been avoided. Conversely, if the decision makers ship off the supplies and then the pandemic arrives full force in their own country, they will later face legitimate questions about why they were more solicitous of the needs of people in other countries than they were for the very people for whose welfare they were responsible. (Complicating all of this is yet another consideration: that sending the scarce treatments to low-resource countries at the epicenter of a pandemic could well be a very prudent move, in the sense that sending the treatments might contain or substantially slow down the pandemic, which in turn could end up saving more lives in the donor country—and perhaps overall—than holding back would have done.)

Recognizing Relevant Ethical Principles

Discussions on these four topics by WHO and its consultants[2] have produced agreement on several basic principles that should underlie planning for and responses to a pandemic. The first is the principle of utility, that is, acting so as to produce the greatest good. One standard criticism of this criterion is that it can lead to a preference for a program that brings very great good to a small number of people, even if that good is not fairly distributed, over a program that brings a much smaller good to a much larger group of people who are well distributed across a society.

A second principle is efficiency, which calls for minimizing the resources needed to produce a particular result or maximizing the result that can be produced from a particular set of resources. The third principle is the principle of fairness, which is usually formulated in a formal manner as treating like cases alike. In this case, I would think that the principle of fairness should deal specifically with the risk of unfair discrimination, that is, discrimination based on irrelevant or illegitimate characteristics of a person or a group.

Finally, there is the principle of liberty, which holds that one should impose the least burden on personal self-determination that is necessary to achieve a legitimate goal—or, in other words, one should not trade all freedom for security.

Additionally, there is a set of principles concerned with the procedures by which decisions are made. The first of these principles is often referred to these days as transparency. It states that information about the processes and bases of decisions should be made available to the affected population. But it isn't enough that things be transparent. There is also a principle of participation—that is, that the stakeholders should be involved, through appropriate institutions and means, in the processes of formulating the objectives and adopting the policies. (The

---

[2]In our pandemic influenza work at WHO, each of these four topics is being addressed by a working group, which met once together (in May 2006) but which mostly have been operating separately as virtual committees. The working groups' papers were used as background for a global consultation in October 2006. The material presented represents an unofficial distillation of the working groups' analysis and not a position of WHO.

efforts of the Pan American Health Organization toward that sort of participatory policy formulation were described in Chapter 2.)

Then there is the principle of review and revisability. Stakeholders should have a way to appeal policies after they have been adopted, and processes should be in place that allow policies and plans to be reviewed and revised in light of experience.

Finally, the principle of effectiveness states that there must be ways to translate the other principles into practice. Otherwise, the principles will have no relevance, and the whole exercise of judging the ethics of pandemic preparation and response becomes irrelevant.

Now, if these eight principles strike you as straightforward and commonsensical, I concur. Indeed, whenever the public is involved in a process, it is important to have some sort of guidelines like these that are readily accessible, that coincide with people's general sense about the way their lives should be lived, that guarantee people an opportunity to participate, and that affirm to people that they will be treated fairly and with respect. In sum, these eight principles in some sense summarize people's general expectations of how government should respond to problems, including problems facing public health authorities. And it is significant that the relevant principles do not apply just to professionals in the private sector but also encompass public health actions involving governments.

Ethical Issues in Access to Health-Care Services

Returning to the four areas of ethical concern, we will now examine some of the specific issues that arise in each of the four areas and see how the above ethical principles apply.

Let us begin with the issue of access to health-care services. The central problem here is how to fairly distribute health-care resources that are not going to be adequate to provide for everyone in need—even in rich countries—if an influenza pandemic occurs anytime soon. Whether it is the United States not yet having anywhere near an adequate stock of Tamiflu, the inability of many countries—even countries of the North—to rapidly supply adequate amounts of vaccine, or the absence in poor countries of even the basic rudiments of "health for all," every country will face a pronounced scarcity of what may be essential, life-saving resources. And it is important to keep in mind that, as has been mentioned already, the scarcity will encompass not just the antiviral drugs and vaccines needed to battle the pandemic but also supplies for day-to-day health care, including emergency and routine surgery, intensive care, and primary care.

In countries where the health-care system is hanging on by its fingertips, if at all, the notion of spending a lot of time on planning for pandemic influenza itself raises ethical problems. Such preparations are probably important, but choosing to undertake them means that a country is immediately facing a trade-off, as specialists will be drawn away from activities that are necessary for care and planning that is immediately relevant today to plan for a contingency that may not occur for quite some time, if ever.

At the heart of all of these considerations is the question: What counts as a fair distribution? The answer to this question in turn rests on what one thinks about two questions regarding justice.

The first question is: What sort of justice is being sought? If we are looking for compensatory justice, that means we believe it is important to make up for any special burdens that a person has suffered. The burdens in question could arise as part of a pandemic response in a number of different ways—for example, when a poor subsistence farmer is forced to get rid of his small number of chickens because somewhere within a few kilometers a case of avian influenza has been detected.

Another approach is generally referred to as distributive justice. Among people in a society there are those who are generally worse off than others, and some believe that any time a government undertakes efforts to improve society—as, for example, through public health activities—it should try to make people in the disadvantaged group relatively better off. This is an idea that is often associated with the philosopher John Rawls, who thought it important to attend first to the needs of those who are worst off in society. That statement, of course, raises yet another question: Worse off in what respect? Worse off in terms of health, in terms of income, or just what should we be looking at?

Then there is procedural justice. Philosophers such as Norman Daniels have written a good deal about the characteristics of fair process, which is an aspect of justice that is particularly relevant in situations where it is difficult to achieve wide social agreement on the substantive implications. In such circumstances, if the process used to reach a decision is perceived as being fair in the way that it treats the interests and views of people, then the results should be seen as more justified than if the process was not seen as a fair one. And, as a practical matter, the results of such a decision may also be more widely acceptable even if the decision itself is not popular with many people.

The second basic question regarding justice is: What is the basis of comparison? Again, I have already spoken to the issue: Are we looking overall for the maximization of well-being? Is it important to us that the effects on well-being are comparable among different members of society? Are we concerned only with lives saved? Are we concerned with achieving justice in terms of economic and social costs? To see how these various principles might play out in the face of pandemic influenza, in the next section we analyze in more detail what happens to society in the course of such a pandemic.

The Issues

The health-care access issue with the highest visibility in terms of the attention that it has gotten in the press is, I believe, access to Tamiflu®. In particular, the issue of prioritization raises a number of difficult questions: Should the Tamiflu® go first to health-care workers, and, if so, does that mean just to the professionals or also to others working in the hospitals, perhaps even including the cleaning staff? Should it go to other first-responders preferentially and others who provide public services? And which services? Should preference be given to those whose service involves maintaining safety in society, such as police officers and soldiers? How about those who provide educational services? How about those who deliver everything from medical supplies to groceries? If truck drivers are going to be too sick to drive across the George Washington Bridge or go through the Holland Tunnel, how quickly will the stores of Manhattan run out of food?

Should priority be given to those most likely to be stricken and to die? Those most likely to survive if treated? Those most likely to spread the disease if not treated? Should all lives be valued equally, or should there be a preference for saving those most productive for society, which would generally imply young to middle-aged workers? Or should there be a preference for the youngest based on an argument sometimes referred to as "fair innings," namely that older persons—say 65 years and older—have already had their fair share of opportunities and that persons of 15 or 25 deserve a chance to have theirs.

To illustrate how these factors might actually play out, let me share with you an exercise developed for the Massachusetts Health Department by some of my colleagues at the Harvard schools of medicine and public health, principally Professor Dan Brock. Assume that for the past several months there has been sustained human-to-human transmission of a novel strain of avian influenza A with genetic components of human influenza in several countries around the world. Your community was first affected three weeks ago, and, since then, there have been over 500 cases and 50 deaths.

Tamiflu is the only drug that may effectively reduce mortality of ill patients and limit infection of exposed persons. However, supplies are limited, and hospitals across the country are independently making decisions about how to govern the allocation of such antivirals within their own institution. In your community, the four major medical centers have recently established four different protocols regarding prioritization of access to care:

- Hospital A, recognizing the importance of protecting its work force in order to minimize absenteeism and ensure continuous response capacity, has decided to use its remaining cache of Tamiflu® for prophylaxis of staff who are exposed while caring for influenza patients.
- Hospital B, in an effort to save its very ill patients, has decided to reserve its remaining Tamiflu® for treatment of the sickest influenza patients. This approach is consistent with the usual practice of providers at hospital B, who are accustomed to focusing primarily on treatment. Hospital B is relying on airborne-infection isolation and personal protective equipment—namely, N95

respirators, gloves, and gowns—to protect its staff, and is not using Tamiflu® for prophylaxis.

- Hospital C, in order to maximize survival rates, has decided to reserve its remaining Tamiflu® for treatment of those patients most likely to benefit, namely, those who come to the hospital within 48 hours of disease onset. As this prioritization plan will result in faster depletion of the antivirals, hospital C is relying on airborne infection isolation and personal protective equipment to protect its staff and is not using Tamiflu® for prophylaxis.
- Hospital D, assuming that its cache of Tamiflu® will soon be depleted regardless of its distribution strategy, is using the antiviral for prophylaxis of exposed staff and treatment of all probable and confirmed cases, regardless of severity. This is the most comprehensive approach, and hospital D will reach the limits of its available Tamiflu® stock most quickly.

Professor Brock has provided a set of five questions with which to analyze the different strategies:

1. Is each hospital's plan independently something that would be fair and reasonable? Viewed in the context of the community, what are some potential challenges that may arise as the result of different institutions using these different strategies?
2. Are these strategies publicly announced? If so, it will not only give each hospital a reputation as to what its ethical orientation is, but it will obviously lead everyone in the community, from hospital workers (if they are able to work at more than one place) to the very sick and to recently exposed patients, to choose one hospital or another based on how their own perceived needs fit with the announced policy.
3. Which of the options should be employed when prioritizing the allocation of limited resources?
4. What factors should govern this decision? Should this decision be made by leaders within the individual hospitals, by local government, by public health officials, by the state department of public health, or perhaps by the national government? Should there be one uniform policy for all?
5. What are some prospective actions the community could take to avoid reaching this point? Who should be involved in this process? Who should make the decisions? Should the state department of public health mandate that hospitals unify their actions?

Another allocation issue involves the distribution of risks rather than benefits. This issue arises, for instance, regarding the subjects in the clinical trials of new vaccines or drugs. Suppose there is a push to accelerate the technical and ethical review process, not simply to get the regular process done more quickly but maybe to omit certain time-consuming items because they do not seem to important enough to worry about in the face of an impending pandemic. The issue also arises when considering which clinical interventions to use to fight a pandemic. Is it acceptable to approve interventions under interpretations of standards that differ from the usual interpretations, when doing so increases the risk of lack of efficacy or, worse, of actual harm for people taking the approved product? Consider, for example, the case of an adjuvant that is not usually approved in the United States for use in vaccines. Would it be acceptable to use that adjuvant in a vaccine if that would increase number of doses, assuming that doing so would not involve at most a modest increase in risk of harm or reduced efficacy?

A third issue of risk-allocation is whether health-care workers are obliged to accept vaccination. There are at least two arguments for why the answer should be yes. One of them parallels the argument used to justify members of combat divisions being made to undergo mandatory vaccination against possible agents of biowarfare so as to keep them fit for battle, even when the vaccines are newly developed or perhaps experimental. A second argument goes beyond that and focuses on the need to protect the patients with whom health-care workers come into contact. And this, of course, takes us back to the reciprocal issue, namely, the providing of benefits to frontline workers because they run a disproportionate share of risks.

Arriving at Ethical Policies for a Pandemic on Access to Health Care

What conclusions might one reach about how treatments should be allocated during a pandemic? First, it is very clear from the above four-hospital example that the fairness of the process will be crucial in producing allocation decisions that are defensible. After all, none of the alternatives adopted by the hypothetical hospitals A, B, C, or D is prima facie unreasonable, and no one choice is clearly the correct one, so the legitimacy of the process is going to be very important. This means that information, participation, transparency and revisability will be key to the decision-making process: How are the priorities being set, and who has a say in that process? The standards need to be publicly articulated, debated, and justified in light of objectives that are either agreed upon or are themselves at least openly debated.

My suspicion is that the greatest cause of skepticism about the fairness of the process will be policy-setting processes that are opaque and exclusionary. This would be true in the United States, but it will be particularly the case in countries with a history of discrimination against certain ethnic or religious groups and in countries where there are tribal differences and where there are suspicions that the government acts on behalf of one group at the expense of another. Ironically, the decision-makers who try to impose their own agenda by excluding others from the process will be the least likely to achieve the goals that they want because policies in this arena are going to be very dependent on community acceptance for their success.

Ethical and Legal Issues in Public Health Interventions

The second area of ethical concern involves public health interventions, such as the quarantine of exposed persons, isolation of infected persons, social distancing among the general population, border control, personal hygiene, and so forth. One of the ethical questions that has to be asked is whether we actually know how to measure the benefits that will come from any of these interventions. This is a particularly important issue if one is seeking to balance limitations on personal freedom against the value of achieving a legitimate public goal that cannot be achieved in any less intrusive or less burdensome way. If we cannot measure the expected benefits—that is, if we do not have a data-based model showing that performing a particular action will save this many lives or prevent this many illnesses or save this much money that can be used for other purposes, or whatever the benefit might be—then imposing on people's right of self-determination would not fit within the ethical framework that I articulated earlier.

It is clear that the specific decision-making process is going to be very important here, perhaps even more so than in the allocation area. As interventions and burdens are imposed on the public, it will become very important that the process by which people's freedom is limited has been publicly laid out, so that people can participate in that process if they choose. Just as different groups may make different allocation choices, different communities and different people may have different levels of risk aversion and may put varying weights on the value of preserving liberty. There is an intersection here with the allocation issue discussed above because one of the limitations on liberty could be the imposition of public control over drugs and vaccines to meet whatever plan has been decided upon to maximize public benefit, rather than leaving choices about drugs and vaccines to individuals. In selecting a public health intervention, one should always seek the least restrictive alternative, as chosen by people with authority to act on solid evidence.

Obligations of and to Health-Care Workers

As I have already suggested, the unique skills of health-care workers create certain prima facie obligations on them to provide needed care based on one (or both) of two arguments: first, that persons who enjoy a privileged position are obligated to respond when the society that bestowed those privileges is in great need; and, second, that health-care skills are established and judged by self-governing professionals, and this status has long been understood to carry with it certain ethical obligations toward patients and the community, as recognized in the Hippocratic Oath and other professional codes.

The question remains, however: What are the limits of this obligation? The level of risk that individuals are willing to accept is a matter of personal choice, but choices that depart from recognized obligations may be subject to sanctions.

A number of factors are important here, including how the disease is actually spread—and, in particular, whether health-care workers are at heightened risk—and the obligations that health-care workers have to their families and significant others. The issues are always what obligations these people have and how they compare with the obligations of non-professionals who may also be placed at increased risk.

Linked to this are society's obligations to health-care workers if it turns out that the disease is not being passed simply by community exposure, where everybody is at equal risk, but instead its spread is closer to that of SARS or Ebola, where the activities that health-care workers, including public health officials, undertake place them at higher risk than other members of the community. If providing care increases risk for health-care workers, there is a legitimate reason to provide them with preferential access to prophylaxis and treatment on the grounds that are narrower than simply rewarding general social utility—that is, using scarce medical resources on health-care workers would be directly associated with the continued ability of the health-care system to provide these interventions to everybody else. This is not true of many other people who provide social benefit to society, but it may well be true of some people who are not usually described as health-care workers (e.g., people who deliver health-care supplies).

There are two ways of looking at this. The first is that if there is added risk, the choice to take it on should be, at first, a voluntary choice. Choosing the risk is therefore a superogatory—and hence particularly praiseworthy—act. The alternative approach—in my mind, less preferable, though it may be necessary—would be for health-care workers to be obligated to act because their conduct is mandated by the profession or by the state.

Planners need to work with local and national professional associations in advance—right now—to obtain agreement within these groups about what professional ethics implies about the actions that will be expected from a particular professional in the case of a pandemic. These statements should be developed and agreed upon through a process that is transparent, and there should be opportunities for professionals to hear from the general community about its expectations. In particular, if any sanctions will be applied, that must be clear in advance.

Ethical Aspects of International and Intergovernmental Obligations

National and international issues are clearly interrelated, not the least because the principal international response is going to come from national governments. The impact of a pandemic will be global, so international action will be needed. International disease surveillance is being organized under the new International Health Regulations (IHR), which WHO member states are already implementing voluntarily in advance of the June 2007 deadline when the IHR will formally come into effect. Furthermore, expert advice from WHO on various standards—

including ethical standards—is providing a framework for an internationally coordinated response.

A central purpose of the United Nations, including its specialized agencies such as WHO, is to achieve international cooperation in solving international problems of a humanitarian character. The Universal Declaration of Human Rights guarantees respect for economic, social, and cultural rights, which include matters of health that are indispensable for human dignity, and it proclaims that these rights should be realized through national effort and international cooperation. The International Covenant on Economic, Social and Cultural Rights (to which not all countries are signatories, though I need not mention which ones) commits each state to taking steps—individually and through international assistance and cooperation, especially economic and technical, to the maximum of its available resources—that are aimed at achieving the full realization of the rights set forth in the covenant, which include the promotion of health and the prevention, treatment, and control of epidemics (Article XII). Similar commitments were reaffirmed more recently in the Millennium Declaration.

Despite these international agreements, difficulties arise regarding all of these commitments. The agreements offer laudable goals rather than the precise steps that need to be taken. They are vague on how they are to be implemented. They are focused on long-term development and on the progressive realization of rights, not on concrete, rapid response to emergencies. They provide no criteria for judging how a state should assess the extent of its obligations or to whom such obligations attach. Obviously, governments are more likely to act if others are acting, so there are always questions of coordination and initiative: Who gets the ball rolling and how is it moved toward the goal?

Of particular concern when implementing human rights obligations is keeping the focus on those persons who are especially vulnerable. This vulnerability is partly determined by biology, but it may also be determined by political circumstances. That is, if you are a citizen of a country that does not have the ability to mount a well-organized response to a pandemic, you are at greater risk than a person in a country in which the virus is equally prevalent but which is better prepared to respond.

Those countries that are considering providing medical supplies, personnel, and other support will naturally be more inclined to do so if the recipient country has in place a plan that makes it likely the donations will be used both efficiently and fairly: Don't ask us for aid if you're just going to take care of the elite, the well-connected, or the armed forces. Yet the obligations that countries have to people in other countries—under human rights conventions and, for that matter, simply on moral grounds—are not primarily obligations to other governments. They are obligations to the people at risk. Although such obligations are typically met by supporting the governments that represent these vulnerable populations, if those governments are unable to act or unwilling to take the needed steps, the humanitarian obligation should, if possible, be met by other means.

Perhaps this situation may provide an additional impetus for countries with the greatest need for assistance in responding to a pandemic to take the necessary steps in planning for a pandemic now. For all societies, the ability to respond appropriately will depend upon being aware of the problem and the threat that it poses, possessing the knowledge, expertise, and technology to control the problem, and having the financial, logistical and administrative capacity to act effectively. (When I speak of the ability of the society, I mean all its components, including voluntary, community, and professional bodies, not merely central or regional governments, though they may take the lead in stimulating and directing the planning activities.)

In turn, the obligation of potential donor states—those that have the ability to respond to the threat of a pandemic with needed medical supplies, personnel, funds, and the like—depends on several factors: first, the extent to which their own capacity exceeds a reasonable projection of their own population's needs; second, the existence of an organized international response; and third, the existence in potential recipient countries of plans to use the aid in an effective and equitable manner. Because aid activities are often undertaken in service of donor countries' political or prudential aims, these activities have often been carried out without regard for whether they support a well-planned and ethically defensible response in the recipient country. But when scarce public health resources are being taken from domestic use for the benefit of people elsewhere, it is imperative that these resources be allocated to those places that will use them efficiently and ethically. A country that has that capacity is in a much better position to argue for scarce aid than a country that does not, not only on the grounds that it has done its planning ethically and has organized itself in a fair way, but also on the grounds that the resources spent there will be more likely to have a beneficial effect.

A Few Concluding Remarks

Having discussed some ethical theories and the different models of pandemic preparedness they might lead to, let me underline some straightforward, practical suggestions that I think most ethicists would agree with. First, the key to an ethically responsible and appropriate response is advanced planning, including communication. Part of the communication is openly acknowledging the unavoidable reality of scarcity of life-preserving resources and thus the resulting need both for collective action and for personal responsibility.

As a practical matter, the likelihood of being able to carry out any pandemic response plan will be greatly enhanced if the general population is aware of and, to the greatest extent possible, involved in the planning process in advance of a crisis. This is particularly true when, as will certainly be the case in a pandemic, success depends not only on acceptance by the public but on their cooperating with requirements that will demand their forbearance or self-denial. If I may draw an analogy, I would suggest that planners need to engage the public in something that, about 50 years ago, Irving Janis termed "the work of worrying." Physicians who are reluctant to have frank conversations with their patients about the problems that may arise in a course of treatment or surgery often express concern that doing so will only makes a patient very anxious, perhaps for nothing since the problems may not actually arise, or worse, that it will induce symptoms precisely because patients are worried about them. Janis's findings about the value of psychological rehearsal of possible problems countered this concern. He found that the work of worrying made people stronger and more capable of dealing with the problems that did occur, while the imagined problems that did not occur produce no long-lasting effects. Likewise, I would argue that public health leaders should be prepared to lead people through the work of worrying about the effects of a pandemic on their lives. It is very difficult to get people to focus on future difficulties that remain merely possibilities, so when the occasion arises to engage people in some needed worrying about what difficulties they will face should a pandemic strike—realistic planning efforts that involve simulations of disasters, for example, or some other event that can seize the public's attention—I believe the occasion should be exploited.

An ethical conclusion related to the international picture is that it is important to recognize the full meaning of "solidarity." We often think of solidarity as most relevant when poor countries make pleas to rich countries for emergency assistance. Yet it is also relevant in the situation when a mutated influenza virus, capable of sustained human-to-human transmission, first appears. The countries first affected are likely to be relatively poor countries, and their governments are expected under international agreements, especially the IHR, to act responsibly by carrying out surveillance, reporting outbreaks to the international community, and supply samples of the infectious agent to the appropriate agency. The correlate of this is that the governments with greater resources ought to make clear in advance—as they have begun to, through pledging funds to the international preparations—how these essential activities will be rewarded through efforts to help the first-affected countries deal with their own health problems and contain the spread of the virus as long as possible (including fair access to vaccines), as well as through financial aid to mitigate harm caused by disruption of trade, loss of tourism, and the like. If widespread culling of poultry—which could perhaps destroy a major sector of the economy—is needed to keep avian influenza from moving into the human population, what reciprocal duties fall on those countries that stand to benefit from this measure? Likewise, what if air travel or shipping must be suspended? The countries that benefit from such efforts have an obligation to help compensate the frontline countries.

As this set of issues illustrates, our entire topic can be seen through an ethical lens as the need to resolve the classic struggle between individual and group. Here we have, for example, the interests of country A versus the interests of the community of nations. We have the interests of one person in obtaining scarce resources to protect or treat himself or herself (or family)—resources for which the person may be willing and able to pay—versus the interests of others in having those resources distributed according to established priorities. We have the interest of a subject in a clinical trial to have that trial conducted cautiously versus the interests of people who are waiting to receive needed drugs or vaccines and wish to have them tested and approved as quickly as possible. We have the interests of physicians and other health-care workers in being able to protect themselves from influenza competing with the needs of the group to have functional health-care services respond as fully as possible to a pandemic. And we have the interests of an individual to move freely and to decide about his or her own treatment versus the interests of the community in isolating or treating or vaccinating people who could spread a deadly disease.

All of these ethical dilemmas are dilemmas precisely because they involve choices between one good and another good. Unlike many human-rights situations, they do not involve choosing between doing the right thing, on the one hand, and clearly violating people's rights, on the other. They are difficult because both choices are good ones, and they seem particularly difficult because of our emphasis—in Western societies generally and in health care particularly—on respect for individual choices. The collective interests embodied in traditional public health measures are less often acknowledged despite their clear and strong history.

So it becomes very important for ethical as well as practical reasons that the decision-making processes involve open public debate. Even if this leads to different outcomes in different communities—and it will certainly lead to different outcomes in different countries—it can provide a basis for ethically justifiable decisions, and the decisions will be ethically justifiable not because they are incontrovertible but because a process involving public communication and participation promotes understanding and acceptance and, therefore, a greater ability to cope with public health measures that are inevitably going to be imperfect. Regrettably, the countries least inclined to engage in such processes and least capable of doing so may be precisely those where the need is the greatest. And so for ethical as well as prudential reasons, part of the U.S. response to the threat of pandemic influenza should be to find means of helping people in other countries, as well as those here at home, to engage in the work of worrying that is an essential part of our individual and collective pandemic preparations.

SOCIAL JUSTICE AND PANDEMIC PLANNING AND RESPONSE

Ruth Faden, Ph.D., M.P.H.[3]
Johns Hopkins Berman Institute of Bioethics

It is reasonable to expect that the health, economic, and social burdens of a pandemic influenza will fall disproportionately on the poorest countries of the world and on the poor and otherwise systematically disadvantaged within the world's rich countries. Arguably, the greatest moral challenge posed by a pandemic is how to respect commitments to social justice in the face of the overwhelming and entrenched inequalities in health, well-being, and resources that will constitute the backdrop for, and the harsh realities of, any global outbreak of devastating disease.

As Madison Powers and I argue in our recent book, *Social Justice*, at least three moral tasks are required to address this challenge (Powers and Faden, 2006). The first task is to identify which of the likely inequalities in burden are the most morally egregious. Not all the ways in which some will suffer less and others will suffer more are ethically problematic to the same degree, so it is important to identify which of these inequalities are the most unjust.

The second task is to determine, among the more egregious inequalities, which are the most easily prevented or at least mitigated. Although, tragically, there may be neither the political will nor the economic capability to address some of the likely egregious injustices, by focusing on those that are the most plausible candidates for public policy interventions or technological fixes, it may be possible to at least blunt or narrow some of the worst injustices that might occur.

The third task is to implement whatever changes in policy and practice are necessary to reduce those inequalities that are amenable to prevention or mitigation.

In this paper I set the stage for the first of these three tasks by raising some relevant concerns for the United States as well as from a global standpoint. I close by briefly describing a recent effort to address the third of these tasks by the Bellagio Group. The Bellagio Group is advocating that international institutions and national governments adopt a series of principles and action steps to protect the interests of systematically disadvantaged groups in pandemic planning and response.

[3]Philip Franklin Wagley Professor of Biomedical Ethics and Executive Director of the Johns Hopkins Berman Bioethics of Institute.

Social Justice in the Global Context

Recent work is beginning to provide a formal evidence base supporting the assumption that the burden of disease from a pandemic influenza will fall disproportionately on the developing world (Murray et al., 2006; Cummings, in press). Even without the formal evidence base, however, the case for this claim, as well as for the broader assumption that economic and social burdens will be far greater in the developing world, would seem obvious. There is simply no reason to think that the global impact of a pandemic influenza will be distributed any differently than the burdens of other infectious diseases, such as HIV/AIDS, which have uniformly taken a much greater toll on the poor nations of the world.

What should be emphasized most perhaps is the magnitude of the disparity that can be anticipated. Historical data suggest that in 1918 the mortality rate from influenza in the United States and the United Kingdom was in the order of 4 per 1,000 (Collins et al., 1930; Registrar General of the United Kingdom, 1920). By contrast, the mortality rate is estimated to have been between 32 and 55 per 1,000 in India (Mills, 1986) and between 19 and 51 per 1,000 in South Africa (Phillips, 1990). There is little reason to think that this dramatic gap in disease burden will be any smaller in the next pandemic. Disparities in life expectancy and access to primary medical care between the wealthy and poor nations of the world have, if anything, increased in the years since 1918. Although vaccines, antivirals, ventilators, and other advances in medical technology may be able to reduce the burdens of a twenty-first-century pandemic in the world's wealthier nations, it is likely that many people living in the world's poorest regions will have little or no access to these interventions, and thus that the gap in burden of disease may be even greater in a new pandemic than it was in 1918.

Elsewhere I have argued that it is difficult to conceive of a more egregious injustice than the extraordinary gap in life expectancy and well-being that continues to persist between the world's desperately poor and the rest of us (Powers and Faden, 2006). In some respects, the additional hit that the world's poorest will likely receive from a new pandemic is simply more of the same. How directing global resources to narrow the gap in impact of a potential influenza pandemic compares with other strategies to improve the well-being of the world's poorest people is an open—and important—question. In particular, whether pandemic influenza's claim for our moral attention is greater or lesser than the claims of other factors that contribute to the horrible conditions and reduced life prospects of the world's poor depends largely on causal claims about the relative impacts of the different factors and on technical claims about the prospects of successfully intervening in a pandemic.

There are, however, at least two sets of arguments that suggest there may be something morally special about a pandemic influenza. The first set of arguments has to do with some particular features that are not unique to pandemics but that may have particular moral resonance. An influenza pandemic is a discrete threat that is foreseen, rather than an existing state of affairs with a complicated and contested history. Every country and every individual in the world is vulnerable to this threat. Indeed, the term pandemic is used only because it is expected that the disease will be a global phenomenon. This universal threat has rightly prompted a significant response from almost all of the major global institutions, with a high (if not high enough) degree of international cooperation. In the face of this remarkable amount of anticipatory planning and international cooperative action, a failure by the global order to take steps to mitigate the completely foreseeable, disproportionate impact on the world's poor would be a singular moral failing. The global order would be responding to a discrete, highly consequential threat to well-being, to which all peoples of the world were vulnerable, by knowingly taking actions that disproportionately protected those of us who already were relatively advantaged.

A second set of reasons has to do with the fact that the burdens of efforts to prevent or contain a human pandemic, mainly through control of avian disease and surveillance of human disease, are currently falling primarily on the poor countries of the world, while the benefits of these efforts are likely to be experienced mostly by the world's wealthier nations. This imbalance introduces another moral dimension that argues even more strongly that social justice demands a global response to narrow the gap in the burden of disease and in well-being more broadly.

There is increasing evidence that interventions to contain avian disease, particularly through the culling of poultry, are placing a significant and disproportionate burden not just on developing countries but particularly on the poor and low-income people within those countries. While industrial-scale poultry producers are also sustaining losses, they are well positioned to adjust their production practices to changing market conditions. By contrast, small-scale poultry farmers who are just beginning to move up the development ladder have in some cases lost their livelihoods and been plunged back into poverty when their birds have been culled. In some parts of the world, household chickens represent the only source of independent income for poor women and children. When chickens in these villages are culled, the most dependent and vulnerable members of the community become even more dependent and vulnerable. Families and individuals who have been implicated in avian disease outbreaks or in suspected human outbreaks have suffered extreme social stigma and isolation, and there have been reports of suicides and disappearances. Although some of the negative consequences of avian disease control and human surveillance actions could be prevented by better practices, poorer countries—and the poorest within these countries—will continue to suffer the most from containment efforts. Yet if these efforts fail, they again will suffer the most, this time from the pandemic itself.

Social Justice Within the United States

Social justice will not only be the biggest moral challenge of a pandemic across nations, but it will also be the biggest moral challenge within countries. All nations, from the most resource-poor to the wealthiest, are making policies and taking actions that have profound implications for social justice within their boundaries. Whether the rights and interests of the poor, of ethnic and political minorities, of women, and of other disadvantaged groups are adequately taken into account and respected in pandemic planning and response is a central, but ofttimes neglected, question.

Within the Unites States as well as globally, it is reasonable to expect that the health, economic, and social burdens of pandemic influenza will fall disproportionately on the most disadvantaged people. Limited historical data suggest that in 1918 those at the bottom of the American economic ladder suffered the most. For example, in one study of ten locations in the United States, the mortality rate for people classified as "well to do" was 3.8 per 1,000, compared to 10 per 1,000 among those classified as "very poor" (Sydenstricker, 1931). The association between poverty, lowered life expectancy, and increased burden of disease continues to persist in the United States, and there is little reason to think that a new pandemic would defy this pattern. Indeed, insofar as medical countermeasures such as antivirals, vaccines, and ventilators end up mitigating the toll of a twenty-first-century pandemic, it is possible that those at the bottom will have an even greater relative disadvantage than they did in 1918, when these measures were not available. Moreover, it is likely that the poor in our country as well as people who are otherwise substantially disadvantaged will have more difficulty not only in accessing medical countermeasures but also in implementing—and thus in benefiting from—traditional public health measures. In addition, it is likely that the burdens of complying with these measures will be greater for those in our society who already have the least (Blendon et al., 2006).

Consider, for example, the prospect of school closings. While this would be a hardship for many families, for those children for whom school is the safest and most nurturing environment they experience, conditions will be particularly bleak. There will be, at best, no books, no access to the Internet, and no home-based learning. At worst, there will be overcrowding, increased exposure to violence, little adult supervision, and reduced access to healthy foods, if not outright malnutrition.

Consider also the implications of community confinement, isolation, or sheltering in place. People who live paycheck to paycheck could lose what little economic stability they have by the resultant loss of income. Some would have no wherewithal to stockpile food. People who are in and out of housing would have no place in which to shelter.

It is predicted that in the next pandemic influenza there will be an acute shortage of hospital beds. How will those who are profoundly disadvantaged even before a pandemic strikes provide adequate care at home to sick loved ones? Families are not equally positioned to be able to take care of a seriously ill person. Even accessing medical advice in one's home through the Internet or by phone will be more difficult, if not impossible, for some families.

The difficulties that the poor and other disadvantaged groups face in putting public health recommendations into action and the increased burdens that some of these recommendations will impose on them have deeply troubling implications for social justice. Arguably, the moral authority of the government to make such recommendations carries with it the reciprocal moral responsibility to identify where social injustices with respect to these recommendations are likely to occur and to take reasonable steps to reduce, if not prevent, at least the most egregious injustices among them. A pandemic will produce suffering, and suffering is itself sufficient cause for moral concern. But from a moral point of view, of even deeper concern is the suffering associated with an injustice that could have been anticipated and ameliorated.

Similar concerns about social justice apply to medical countermeasures as well. Imagine, for instance, that local governments have vaccine to distribute. In order to effect an efficient distribution of the vaccine, the population is asked to remain calm in their homes and await notice about when to appear at a particular location for vaccination. But in the aftermath of Katrina, why should people who belong to systematically disadvantaged groups trust that the government will distribute vaccine as quickly and in the same quantity in their neighborhoods as in the neighborhoods of more advantaged Americans? It is in the interests of public health and public order, as well as of justice, that disadvantaged groups not be further disadvantaged by how medical countermeasures are distributed and also that these groups do not perceive themselves as being treated unjustly. Reducing perceptions of injustice may be as difficult to do as reducing injustice itself. Most would agree that the worst injustices would involve having only the powerful, the elite, or the affluent receive—or else be the first to receive—whatever is in short supply, whether vaccines, antivirals, intensive care unit (ICU) beds, respirators, or admissions to hospital. Such an outcome would constitute an egregious inequality that should be avoided at all costs. At the same time, it will be difficult to make priority decisions that do not give the reasonable appearance of unjustly favoring the privileged at the expense of the systematically disadvantaged. Consider two examples.

Many universities with sizable undergraduate programs also operate hospitals located in impoverished, inner-city neighborhoods. In the event of a pandemic, undergraduates may be asked, or told, to remain in their dormitories. It is likely that some of these undergraduates will become ill. University officials may feel—and indeed may have—a special moral obligation to care for the young people in their charge. Certainly parents will have a reasonable expectation that the university will use the medical resources at its disposal to assist their desperately ill child when they cannot. It is an open moral question whether, in a context of extreme scarcity, university leaders should prioritize the students in their care over others, including other young people who live in the surrounding neighborhoods to whom the university also has an obligation. But even if such a priority is morally acceptable, or even if it is morally required, it will be difficult to persuade poor parents of color that the bypassing of their children in favor of university students is not a grievous injustice in which the lives of the privileged are valued more than the lives of the disadvantaged.

Turning from the university hospital to public policy, on the subject of access to medical countermeasures many current national planning documents as well as many state and local plans give priority to those people who are deemed essential to maintaining the infrastructures for national defense, public safety, governance, communications, and commerce as well as for public health and medical care. Although there is considerable disagreement about who exactly should be considered essential in each of these sectors, there is widespread agreement that such a prioritization is in the public interest and is morally justified. The problem from the standpoint of people who are at the bottom of society, however, is that they are likely to be underrepresented in the ranks of occupations deemed critical to essential infrastructures. Although the critical workforce will no doubt be more diverse in social class and ethnicity than traditional political or economic elites, it may still be reasonable for those who are most disadvantaged to conclude that their interests and rights are being disregarded in egregiously unjust ways.

The Bellagio Principles

In an intensive meeting held in Bellagio, Italy, in July 2006, twenty-four officials, scientists, and public health and policy experts from eleven countries discussed the rights and interests of systematically disadvantaged groups in pandemic planning and response. This group, which I helped convene with my colleague Ruth Karron, concluded that international institutions and nation-states must bring these rights and interests to the forefront of consideration in efforts to prevent a pandemic, in responses during a pandemic, and in efforts to redress economic and health burdens in the post-pandemic period. The Bellagio Group did not conclude that the rights and interests of disadvantaged groups should always be given priority, but instead that they should always be taken into account as part of a serious commitment to social justice. The Bellagio Group is advocating the adoption of its Statement of Principles (Bellagio Group, 2006a), which is intended to guide institutions and governments wishing to support this commitment. The Statement of Principles addresses the need to reduce the technology gap that currently exists between poor and wealthy nations with respect not only to biomedical research and development but also to relevant socioeconomic analysis. Most of the research that has been conducted so far has focused on the implications of alternative containment strategies for the developed world (Ferguson et al., 2006), including studies projecting the implications of efforts to prevent animal or human disease in Asian Rim countries (Ferguson et al., 2005; Longini et al., 2005). While wealthy nations thus can take advantage of an increasing body of relevant modeling and epidemiological analyses in revising their national plans, poor countries have almost no such research on which to rely.

The Statement of Principles also calls on countries and regions to explicitly identify how disadvantaged groups in their societies will fare under proposed pandemic plans, with respect to both the prospects for benefit and the burdens and secondary harms that pandemic planning and response produces. Whether in tabletop exercises, community exercises, or mathematical modeling, we will not uncover what is likely to happen to those who are already the most disadvantaged unless we make it a priority to find out. Similarly, the Principles call for ensuring that special efforts are made in communications and public engagement to involve disadvantaged groups. These groups are frequently underrepresented in political and civic processes and may doubt the credibility and trustworthiness of those responsible for pandemic planning.

To facilitate the translation of the Statement of Principles into policy and practice, the Bellagio Group developed a series of checklists intended to guide the actions of government officials in developing and revising pandemic-response plans and of animal health and public health practitioners in responding to suspected outbreaks of animal and human disease (Bellagio Group, 2006b). These checklists are being adopted for use in various settings throughout the world. It remains an open question whether they will contribute in any way to reducing at least the most egregious social injustices that are currently occurring in attempts to prevent a pandemic and that will occur should a pandemic take place. There is no question that concerted efforts are urgently needed, both globally and within the United States, to address social justice concerns in pandemic planning and response.

REDUCING STATE VARIABILITY IN HEALTH EMERGENCY PREPAREDNESS THROUGH FEDERAL STANDARDS, ENFORCEMENT, AND PUBLIC ACCOUNTABILITY: LESSONS FROM THE ENVIRONMENTAL FIELD

Shelley A. Hearne, Dr.P.H.[4]
Johns Hopkins University

Against the background of the 2001 anthrax attacks, Hurricane Katrina, and the looming threat of an influenza pandemic, the public health field is wrestling with how to improve its preparedness for a major health emergency in the most effective and ethical manner. The challenge is not new, however: public health

[4]Visiting Professor, Bloomberg School of Public Health.

faced fundamentally similar issues 30 years ago with the emergence of environmental health threats.

For the majority of the last century, health agencies were responsible for environmental management and regulation. By the end of the late 1960s a series of high-profile events, ranging from Ohio's Cuyahoga River catching on fire to heated Congressional debates about lead in consumer products, raised questions about the public health services' ability to protect citizens from emerging environmental threats. As a result, these responsibilities were predominately shifted from federal and state health agencies to newly formed environmental regulatory agencies (Burke et al., 1997). The U.S. Environmental Protection Agency (EPA) began establishing national standards, rules, and regulations as well as extensive public accountability and engagement with the goal of ensuring that all states provided their citizens with equal protections from toxic contaminants in water, air, foods, and communities.

We are at an eerily similar crossroads with emerging bioterrorism and pandemic threats, which once again raise questions about the ability of the public health field to adequately respond. The insights gained several decades ago from the environmental agencies' efforts to ensure fair and equal protections from exogenous threats can help inform federal health agencies today about ways that they can more ethically and effectively address pandemic threats. In particular, without strengthened oversight and transparency in health emergency preparedness, some responsibilities might be shifted to emergency management or homeland security agencies, just as occurred when environmental threat management was transferred away from public health agencies.

The Landscape: Geographic Variability in Health Emergency Preparedness

With substantial federal investments in public health infrastructure, progress has been achieved in strengthening the nation's ability to respond to a biological, chemical, or radiologic attack. Still, numerous national studies have found wide variations from one state to the next in their capabilities to respond to a major health crisis, such as an influenza pandemic. Concerns include the following (Trust for America's Health, 2006):

- CDC reports that only 56 percent of states have tested their state-wide pandemic plan in the last 12 months;
- Only fifteen states and two cities have received CDC's highest rating for preparedness to receive the country's Strategic National Stockpile, the emergency medical supplies delivered in a major catastrophe;
- Eleven states and the District of Columbia lack sufficient BSL-3 laboratories to test biological agents, according to a survey by the Association of Public Health Laboratories (Trust for America's Health, 2006); and
- Only seven states and three metropolitan areas are part of the nation's

active surveillance system for food-borne diseases, which monitors trends and determines specific foods associated with poisoning incidents (CDC, 2007).

This variation in the states' ability to plan, monitor, and respond to medical or biological emergencies could leave citizens with different levels of protection based on geographic location.

The Need to Ensure Equal Levels of Preparedness for All Citizens

One of the greatest challenges in pandemic preparation is that the public health system is not a single entity but rather a loosely affiliated network of approximately 3,000 federal, state, and local health agencies. Through their police powers, state and local governments have primary responsibility for the health of their citizens (Gostin, 2002). According to the IOM, the federal government "plays a crucial role in protecting and improving the health of the population by providing leadership in setting health goals, policies, and standards, especially through its regulatory powers" (IOM, 2002).

The U.S. Department of Health and Human Services (HHS) is the primary federal health authority. As one of its principal tasks, HHS administers Medicaid, a health insurance program that provides coverage for over 50 million Americans. States are given federal support and guidance and are allowed flexibility in implementing the programs, based on their population's needs and on market opportunities. HHS ensures appropriate state implementation through an extensive structure of rules, regulations, and standards. Some HHS public health agencies, like the Food and Drug Administration, conduct regulatory activities.

The CDC is HHS's lead agency for ensuring the national capacity to respond to health emergencies and for overseeing disease-prevention efforts. The agency's standard approach is to work with state health agencies as equal partners, allowing the states to determine priorities and implement strategies through federally funded initiatives. Often, important decisions about health policy are determined through outside consensus-based organizations. Despite being a significant financial and technical resource for state and local health agencies, the CDC rarely, if ever, directly establishes national standards or public-accountability mechanisms for a state's health-protections efforts.

An Example: The Disease Surveillance Challenge

CDC is the lead agency nationally for collecting and disseminating disease surveillance data. Despite collecting information on four out of every five American deaths, the agency has not established national standards for tracking most chronic diseases (Environmental Health Tracking Project Team, 2000). For example, CDC, through its broad authority and the Birth Defects Prevention Act of 1998, provides funding for some state birth-defect surveillance systems, but it does not provide national reporting guidelines (Erickson, 2000). Because of a lack of standards and sufficient funding, there are wide variations in states' reporting systems which create gaps in timeliness, comparability, and consistency and which compromise investigative capabilities (Trust for America's Health, 2005).

Similar challenges exist for emerging infectious diseases. The federal government has made substantial investments in building and fortifying state and local public health preparedness, including communications, laboratories, and surveillance activities, and this in turn has improved capabilities for tracking infectious diseases. But CDC has been limited in its leadership role in directing the system's ability to respond rapidly to emerging infectious diseases.

For example, in 2003, the nation faced an annual influenza season that appeared highly lethal to children. As the CDC director noted in the heat of the crisis:

> It was difficult to say if this year's childhood death toll from the flu was higher than average, because they don't have detailed, accurate information on the number of children who die every year from influenza specifically because it hasn't been a reportable illness (CDC, 2003).

State disease surveillance is mandated through state legislation or regulations and is submitted to CDC on a voluntary basis. As such, states may vary in their disease reporting. In the midst of the 2003 influenza crisis, which had led to an enormous public demand for information, CDC requested a professional association—the Council on State and Territorial Epidemiologists (CSTE)—to consider recommending pediatric deaths due to influenza as a nationally reportable condition (CDC, 2004). CSTE has the responsibility of providing an annual evaluation of the national reportable conditions list through its state health officer members, with CDC input. As of October, 2004, CSTE established influenza-associated pediatric mortality as a nationally notifiable disease that states are voluntarily requested to report to CDC (Jajosky, 2006). Similarly, after the nation's experts had worried for years about the emergence of novel human influenza strains, in January 2007 CSTE finally issued an interim policy calling on CDC to establish national reporting for novel type A influenza virus infections (CSTE, 2007a).

In times of heightened medical threats, if the nation's health agencies are to exhibit national leadership, they must have more rapid response mechanisms. They can still draw on the advice and counsel of key scientist and constituency groups, but the authority for quick, directive action on nationwide disease reporting should exist on the national level. Such a capability could also help the development of coordinated and compatible syndrome surveillance systems and other related terrorism-preparedness programs that are now being devised by the states.

In contrast to the current state of affairs in disease surveillance, in the environmental area the EPA establishes guidance, standards, and reporting systems as part of its standard operating procedures. State agencies and other interested parties can become involved in these efforts through public commenting processes and advisory councils, but the final determination belongs to the EPA.

Status of Measurable, Verifiable, and Publicly Accountable Standard-Setting for Health Agencies

In emergency response, lack of national leadership can lead to an enormous variability among different geographic regions and subpopulations. Variability in turn can cause unequal and sometimes inadequate protection for citizens during a medical crisis. Unfortunately, HHS and its agencies have taken minimal steps to ensure that states are meeting preparedness expectations.

The CDC Public Health Emergency Preparedness cooperative agreement and the Health Resources Services and Services Administration's (HRSA) National Bioterrorism Hospital Preparedness program have established draft performance standards for states and cities. The various jurisdictions provide CDC and HRSA with self-reported data on progress toward achieving these benchmarks, but neither agency has a system to verify or calibrate these city and state reports. The federal agencies have documented nationwide performance on each benchmark, but they have not produced individual state assessments. In fact, there are no review criteria, penalties, or incentives associated with gauging state performance; states have lost funds only for insufficient expenditures. Agencies will not release performance standard data, including outcomes from plan exercises, on a state-by-state basis.

The federal government has required all states to develop pandemic influenza operational plans as a condition of receiving funding through CDC's public health preparedness cooperative agreement. To date all states have, at a minimum, assembled their draft pandemic plans (CSTE, 2007b). In March 2007 all state plans must be submitted to CDC for review and approval (CDC, 2006). However, the agency has not yet established procedures for reviewing or approving these plans. Most likely, if plans are inadequate, states will be asked to modify them rather than face penalties. In December 2006, the Pandemic and All-Hazards Preparedness Act was signed into law with a mandate for CDC to review and approve plans with options for penalty provisions.

Consequently, jurisdiction performance standards and plans lack any form of verification, enforcement, or public accountability.

Limited Public Engagement

Reflecting lessons they have learned from years of intense environmental conflicts and public distrust, such as accompanied the Love Canal or Times Beach incidents, environmental regulatory agencies have taken significant steps to engage, involve, and communicate with the public more actively (Covello and Sandman, 2001). As part of building public trust and support, agencies routinely include the public in standard setting, advisory boards, and risk communication and also provide broad access to data, including information on decision making and agency activities, such as rates of inspections, enforcements, and permits.

As federal and state health agencies states have initiated preparedness efforts, the level of public engagement has varied widely from place to place. In state pandemic preparedness planning, for instance, some jurisdictions actively sought public comments and participation, while others treated the process and documents as "state secrets." In one state, fewer than twenty-five copies were distributed to partner organizations because of concerns that terrorists might obtain sensitive information (Uhlman, 2005), which in turn sparked a barrage of hearings and newspaper articles about the lack of transparency in the process.

Recent investigations have found that greater public involvement is needed in emergency planning for health crises. In 2004 a New York Academy of Medicine (NYAM) study of 2,500 people found that citizens would not respond to government instructions during a major health crisis, such as a radiologic or biologic attack, because the plans did not reflect their individual interests, health, or family needs (Lasker, 2004). These researchers believe that the lack of involvement of the public in planning is a fundamental flaw in emergency preparedness (Lasker, 2006). A RAND study found that public health officials had less engagement with the public than their counterparts in law enforcement (Lurie et al., 2004). The NYAM study concluded that "people are more likely to follow official instructions when they have a lot of trust in what officials tell them to do and are confident that their community is prepared to meet their needs if a terrorist attack occurs" (Lasker, 2004).

To date, federal health agencies have not required public participation in federal or state planning, in evaluation and standard setting efforts for bioterrorism, or in preparedness initiatives for pandemics or other health emergencies.

Evolutions in Health Agency Standards

CDC's Strategic National Stockpile (SNS) program is the most advanced model within the agency for ensuring that all jurisdictions are meeting preparedness requirements. The SNS is a national repository of vital pharmaceutical and medical equipment, such as antibiotics and respiratory ventilators, which in response to emergencies that threaten mass casualties can be delivered to states within 12 hours. Through cooperative agreements, CDC provides states with funding, technical guidance, and checklists on performance standards.

States' SNS programs are inspected by federal teams who evaluate states' self-assessment of their performance standards. Based on total performance, an overall score is awarded using a traffic-light color-coding scheme. As a program once jointly managed with the Department of Homeland Security, the green/amber/red code (with plus and minus variations) indicates a state's overall readiness to effectively manage the SNS (see Table 4-1). Green indicates the highest

TABLE 4-1 CDC Determination of States' Readiness to Receive and Distribute the Strategic National Stockpile as of October 2006

"Stop Light" Readiness Indicator	Number of States in Category
Green	7
Green Minus	9
Amber Plus	9
Amber	12
Amber Minus	6
Red Plus	7
Red	4

SOURCE: Trust for America's Health (2006).

level of readiness; red the lowest. Starting in 2007, states will be measured on a 100-point scale instead of the color-coding system. Regardless of score, all states will receive the SNS in an emergency if approved by HHS/CDC. The scores are used to stimulate states to be optimally prepared.

While CDC has established measurable and verifiable performance standards, states are given flexibility on how to best achieve operational goals.

Broader Approaches for Standard Setting: Accrediting Health Departments

In *The Future of the Public's Health in the 21st Century*, the IOM determined that "greater accountability is needed on the part of state and local public health agencies with regard to the performance of the core public health functions of assessment, assurance, and policy development and the essential public health service" (IOM, p. 157). Building on these recommendations, public health associations have launched an initiative for creating a voluntary national accreditation program with the goal to "improve and protect the health of the public by advancing the quality and performance of state and local public health departments" (Exploring Accreditation Project, 2006). As currently envisioned, state and local health departments may conduct self-evaluations using national standards, which will then be validated by site visits of peers and public members, who in turn will inform an independent board's accreditation decision. The exploratory group believes that accreditation will lead toward continuous quality improvements as well as enhanced credibility, accountability, and public trust in public health agencies (Lasker, 2004).

Recognizing the need for improved accountability, credibility, and consistency in services and capacity, several states have been conducting accreditation-like efforts for years. A Robert Wood Johnson Foundation report found that there are several leadership states with mandatory accreditation or standard-setting programs for their local health agencies that have data verified by independent review teams and that, in some instances, link funding to performance (Theilen, 2004).

Increased Federalism to Encourage and Ensure All Citizens Are Equally and Adequately Protected

From an ethical standpoint, so as to ensure equal levels of preparedness for all citizens, federal health agencies should play a more directive role in establishing standards and critical requirements for state and local jurisdictions. At a minimum the federal government should set

- measurable standards that are verifiable, impact-oriented, publicly available, and performed on an annual basis;
- requirements for systematic public participation in the pandemic preparedness process, including planning, exercise evaluations, and standard setting; and
- provisions for encouraging state and local health agencies to continuously improve, along with penalties, funding incentives, and public-accountability methods.

If the public health field begins to evolve toward a more federalist construct—with national standards, inspection, enforcement, and accountability measures as part of the strategy options—then we will need to start considering how to take the next steps toward ensuring that standards are met. Again, the environmental field offers lessons that could inform and influence public health's evolution toward federalism. In particular, a number of federalism strategies and powers that have been used to encourage equal protection of the public's health and welfare in the environmental area might also have application for pandemic preparedness. These include:

Funding Penalties or Incentives

The federal government can set conditions on state block grants or cooperative agreements to encourage states to adopt federal regulatory standards (Gostin, 2002). This technique has been used to advance various public health protections, as when the federal government required states to set automobile emission standards as a condition for receiving federal highway transportation funds. In a similar way, achieving pandemic-preparedness standards could be made an explicit requirement for cooperative agreements or other state health programs, such as health-care insurance funding. The preparedness standards could require such things as inspections, public participation, and transparent accountability.

Federal Oversight and Preemption

In the Clean Water Act, the EPA is charged with establishing effluent and water standards for waterways. States can be delegated the authority to administer, permit, and enforce parts of the law, but the EPA still retains oversight responsibilities. Should a state fail to meet federal performance standards, the EPA can remove these responsibilities and take over the state's clean-water regulatory activities. Establishing such structured oversight and responsibilities in the case of pandemic preparedness would require legislative action and could be considered in future preparedness laws.

Public Right to Know

Information can be a powerful persuasive force and has been widely used in the environmental field to encourage state and industrial actions. For instance, the EPA provides the public with standardized data on public water systems and their violations of EPA drinking water regulations (EPA, 2007). Similarly, health agencies have effectively used public reporting of restaurant grades as a mechanism for achieving improved sanitary conditions. The federal government could have national reporting requirements for standardized, quantifiable measures on state preparedness performance and on the status of public health infrastructure (such as laboratory capabilities and capacities, workforce, and communications). This information should be collected, assessed, and made available to the public annually, on a state-by-state basis, as a condition of federal funding. As a matter of ethics, taxpayers in each state have a right to know how well their health and safety is being protected.

Citizen Suits

A wide variety of tools to facilitate enforcement of water standards were included in the Clean Water Act, such as provisions that gave citizens the right to sue to force enforcement of federal and state permits and also the right to sue EPA for failure to perform nondiscretionary regulatory duties. In the case of preparedness planning, citizens currently have no such mechanisms available to them, such as the ability to sue the CDC, and provisions for such mechanisms would have to be specifically legislated into public health laws.

Concluding Thoughts

Just as was the case with environmental pollutants, influenza pandemics and other emerging biological threats know no state boundaries. As the nation acknowledges that federal and state governments are underprepared for dealing with a major health disaster, there is an urgent need for nationwide leadership to build a health emergency-response capacity that will provide equal protections to all citizens, no matter what state they live in. The federal government needs improved mechanisms for encouraging and ensuring the preparedness of different jurisdictions, including verifiable and enforceable state and federal performance standards that require public participation and public accountability. CDC must adopt a more directed federalist role so that, regardless of where one lives, citizens will trust that government plans and instructions can and will save lives.

INTENSIVE CARE UNIT TRIAGE DURING AN INFLUENZA PANDEMIC: THE NEED FOR SPECIFIC CLINICAL GUIDELINES

Bernard Lo, M.D.[5]
University of California, San Francisco

Douglas B. White, M.D.[6]
University of California, San Francisco

During a severe influenza pandemic, a dire shortage of breathing machines—mechanical ventilators—is projected. According to one estimate, a pandemic will require 198 percent of the current supply of ventilators (Bartlett, 2006). If this happens, many people in respiratory failure who need mechanical ventilation in order to survive will not receive it. This grave shortage of ventilators will raise unprecedented allocation dilemmas that ought be addressed before a pandemic strikes.

In a pandemic a public health emergency would be declared, and decisions by individual physicians and patients would be subordinated to public health goals (Gostin, 2000). The objective would no longer be the health of individual citizens but the well-being of the community as a whole. Emergency-preparedness exercises have considered which groups should receive priority for scarce public health preventive resources, such as a vaccine or oseltamivir (Emanuel and Wertheimer, 2006; Gostin, 2006). Similarly, ventilators should be considered a scarce resource to be allocated according to public health guidelines rather than by the decisions of individual physicians and patients.

Guidelines for allocating scarce medical resources during a pandemic will require several levels of specificity. At the broadest level, state public health laws express a general societal agreement that during a public health emergency the decisions of individual physicians and patients will be constrained by public health policies (Gostin, 2000). At the next level of specificity—the level of clinical care decisions—hospitals and physicians need criteria for triaging various patients who need mechanical ventilators when the demand greatly exceeds

[5]Professor of Medicine and Director of the Program in Medical Ethics.

[6]Assistant Adjunct Professor, Program in Medical Ethics.

supply during a pandemic. To minimize overall loss of life during a pandemic, priority should be given to patients who require mechanical ventilation but who are highly likely to survive after only a few days on the ventilator. Finally, at the most specific level, frontline physicians need guidance in implementing these triage priorities in specific clinical cases.

Current ICU Allocation Policies

Shortages of ventilators and beds in the intensive care unit (ICU) currently occur sporadically in U.S. hospitals. When this occurs, beds are allocated on a first-come, first-served basis (Society of Critical Care Medicine Ethics Committee, 1994). Patients who are already in the ICU and who need continued ICU care remain there unless they or their surrogates decide to forego it. Patient autonomy is respected. Similarly, mechanical ventilation may be withdrawn without the agreement of the patient or surrogate only if it is futile. Otherwise, the withdrawal of beneficial care without the agreement of the patient would violate the ethical guidelines of beneficence and fidelity to patients, the latter of which requires physicians to put the best interests of current patients foremost, ahead of any obligations to other patients, future patients, or third parties (Lo, 2005; Beauchamp and Childress, 2001).

The sporadic shortages of ICU beds typically stimulate short-term measures to increase ICU resources. For instance, hospitals close their emergency departments to divert ambulances to other hospitals in the city. In addition, hospitals cancel elective surgery and use post-operative rooms as temporary ICU beds. Furthermore, patients whose need for ICU services is lower may receive their care in general hospital beds—a patient recovering in the ICU may be transferred out of the ICU a day earlier than usual, for example. Such measures may not be feasible during an influenza pandemic, however. In the case of ventilators, for instance, during a pandemic it is likely that other hospitals in the same geographical area will also have shortages of ventilators. (For the sake of argument, this paper will assume that hospitals have already taken steps to maximize the availability of ventilators.)

In other clinical situations, several general ethical principles have been used to decide how to allocate scarce resources (Veatch, 2005; Baker and Strosberg, 1992). As we have discussed, the current method for allocating ICU beds during sporadic shortages is first-come, first-served. A second approach gives priority to those who are sickest and therefore in greatest need. During the system of triage established in the Napoleonic army, for instance, soldiers who were "dangerously wounded" received care before the less severely wounded. The dying were left untreated. This approach is still used in modern emergency departments, where medical need determines which patients are seen first. The ethical guideline underlying both is to help those patients who are the worst off and who have the greatest need. Yet a third approach is to maximize health outcomes in the community by granting priority to patients who can be most efficiently treated. This focus on population-based outcomes was first used by British physicians during typhoid epidemics in the eighteenth century. The U.S. military has also adopted this principle of maximizing the total number of lives saved during extreme battlefield situations (Repine et al., 2005). Because these broad allocation principles are contradictory, it is essential to forge agreement on which will be applied during a pandemic before such a dire shortage occurs.

Ventilator Shortages During a Pandemic

Suppose for the sake of a dramatic example that an ICU in the midst of a pandemic has only one available bed and ventilator. In the emergency department are several patients in respiratory failure, all of whom will die without mechanical ventilation. It is not feasible, given staff shortages, to keep these patients alive by manually squeezing a bag to drive air into the lungs. One patient is a 30-year-old whose only medical problem is respiratory failure, presumably from influenza. Another patient has not only respiratory failure from influenza but also hypotension and renal failure. The presence of these additional problems means that the second patient has a worse prognosis than the first (Graf and Janssens, 2005). Additionally, there are two other patients in the emergency department with respiratory failure who also will die without mechanical ventilation. One is a 22-year-old with an acute asthma attack who has no clinical evidence of influenza. Another is a 58-year-old who requires emergency coronary bypass surgery for continued myocardial ischemia despite optimal medical management. These latter two patients are expected to survive if they receive just a few days of mechanical ventilation. Thus the shortage of ventilators will affect not only patients with influenza but also those who have respiratory failure from other causes.

This scenario dramatizes the dilemmas and tradeoffs that frontline physicians are likely to face during a pandemic. First, the goal of helping those most in need will clash with the goal of trying to minimize deaths. Among patients with respiratory failure, those with multi-organ failure and those whose condition deteriorates over the first few days of treatment have a poor prognosis, so they are most in need. But treating these patients will increase the total number of deaths: treating a patient with multi-organ failure, who will need to use a ventilator for many days, will preclude the treatment of other patients who are likely to need it for only a few days. A second dilemma concerns the distribution of benefits. Allocation policies that are neutral on their face, taking into account only medical prognosis, may in practice have disparate impact on different social groups. For instance, members of ethnic groups who historically have suffered discrimination are more likely to be poor or have poor access to medical care. They may have difficulty coming to the hospital and therefore may be sicker when they present. Thus an allocation policy based on evidence-based determinations of medical prognosis and administered in a fair manner may still result in treating disproportionately fewer patients from disadvantaged ethnic minority groups.

Ethical Guidelines for Triage of Mechanical Ventilators During a Pandemic

A scarcity of ventilators during a pandemic will require an allocation policy based on different ethical guidelines than those governing usual clinical care. The term triage is commonly applied to the process of sorting, classifying, and assigning priority to patients when available medical resources are not sufficient to provide care to all who need it. Although the term has been used to refer to a variety of clinical situations, we use it here rather than the term allocation because it carries the connotation of being used in disasters, such as deadly epidemics or battlefield situations, where the shortage is extreme and people die who might be saved if they had access to the level of medical care available in ordinary circumstances.

The first ethical guideline for ventilator use during a pandemic is that increasing the number of lives saved may take priority over patient autonomy. Public health officials, working in concert with clinical experts and public representatives, should set guidelines for prioritizing patients who need mechanical ventilation. Individual physicians and patients must then make decisions that are consistent with these guidelines.

The second guideline is that patients with a high likelihood of surviving after a few days of mechanical ventilation should receive the highest priority. Characterizing this group will be difficult, however, because data are incomplete and uncertain. The prognostic model that has been most thoroughly studied uses clinical data from the day of admission to the ICU (Knaus, 2002), but other studies suggest that patients who have a positive response to treatment in the first 24 to 48 hours have a better prognosis than patients who fail to show such a favorable early response (Graf and Janssens, 2005). There are no published studies that predict the duration of mechanical ventilation that will be needed. Furthermore, no studies have been carried out during a pandemic, so data will need to be extrapolated to this situation. In light of these gaps in the data, decisions will need to be based on consensus and expert judgments. Reaching such consensuses will require extensive discussions and will need to be done before a pandemic occurs.

The third guideline is that during a public health emergency fairness and perceptions of fairness are crucial (Lo and Katz, 2005). Citizens will be more willing to subordinate their personal self-interest to the common good if they believe that the same rules apply to all. Conversely, people who believe that others are receiving special consideration are less likely to accept mandatory emergency measures. Even the perception that some persons are receiving favoritism may undermine willingness to sacrifice for the sake of the community.

The fourth guideline is that transparency is essential during a public health emergency. The public needs to know how ventilators will be allocated in order to trust that the allocation is fair. Triage priorities and policies should be explicit. The public should have ready access to the triage guidelines, the data and the reasoning underlying them, and the process by which they were derived. Such information could be made available on the Internet, for example. From a practical point of view, the public needs a consistent message. Public health announcements through the media should prepare patients and families for individual discussions in the hospital regarding triage.

Applying Triage Principles to Specific Cases

Even if there is wide agreement on the triage principle of minimizing loss of life during a pandemic, hospitals and health-care workers will still face many difficult decisions when making triage decisions in specific cases. Before a pandemic occurs, it will be important to identify these dilemmas, analyze them, and reach some agreement on how to resolve them.

During Triage, Should Patients Already on Ventilators Be Reassessed?

We have framed the problem of allocating ventilators as "the last bed in the ICU." In reality, the situation is more complex because patients already in the ICU on ventilators may have a worse prognosis than new patients with respiratory failure. Suppose, for example, that one of the ICU patients is a 38-year-old man with influenza who has developed multi-organ failure and whose condition has worsened during five days of intensive care. His prognosis now is worse than that of a new patient who presents with respiratory failure as her only medical problem, with no other organ failure. Or suppose that there is also a 68-year-old patient with chronic emphysema and respiratory failure who is gradually improving but who is likely to require several weeks of ventilator support as his lungs slowly improve. Keeping such current ICU patients on ventilators leaves fewer ventilators available to other patients in respiratory failure, who will die without them and who are likely to survive after receiving ventilation for only a few days. Therefore, allowing patients already in the ICU to remain on ventilators without regard to new patients with respiratory failure is likely to decrease the total number of lives saved. On the other hand, removing patients from ventilators who are not improving after several days would violate the usual ethical guideline that a physician should act in the best interests of patients and be faithful to them. Logically, there is no difference between stopping a ventilator and not starting it in the first place, as long as the justification is the same in both cases—in this case, following emergency public health regulations (Beauchamp and Childress, 2001; Lo, 2005). However, health-care workers and families may find it more difficult emotionally to withdraw a ventilator. These emotions need to be anticipated and addressed.

Who Should Make Triage Decisions Regarding Ventilators?

Separating the roles of triage and clinical care allows physicians who are treating patients to remain loyal to those patients during a pandemic. A senior ICU physician in the hospital can be appointed to make triage decisions, so that treating physicians are not forced to decide to withhold or withdraw mechanical ventilation from patients who still desire it. This creates a situation where the triage officer is making decisions based on the overall outcomes for the population, while the treating physician is free to serve the best interests of the individual patient within the constraints of the public health emergency. But the role of triage officers needs to be specified in some detail. Questions to be addressed include: What training will they receive? What decision support and consultation will be available to them as difficult decisions arise? What emotional support will be available to them?

What Other Considerations Should Be Taken into Account During Triage?

We have identified a high likelihood of survival and a short-term need for mechanical ventilation as two criteria for giving high priority to patients with respiratory failure during a pandemic. If there is still a shortage of ventilators after these criteria have been applied, a number of other criteria might be considered. Such criteria might include the likely duration of life and the likely quality of life in a patient after treatment or the existence of personal behaviors that may have led to the respiratory failure, such as smoking or non-adherence with asthma medications. Judgments about quality of life and personal behaviors are more subjective that a strict medical prognosis and inevitably involve value judgments over which reasonable people may disagree. Because incorporating these considerations into triage decisions would heighten concerns about unfairness, they are best avoided during a public health emergency.

How Will Disagreements by Family Members Be Managed?

Civilians have no experience with triage, unlike military personnel who are familiar with the approach. Faced with the death of a relative which might be averted with mechanical ventilation, families might strongly object to foregoing the use of the ventilator. In light of this, several issues likely to face frontline physicians should be addressed before a pandemic strikes. Would it be feasible, for example, to create timely appeals mechanisms for decisions regarding ventilator use? During public health emergencies, governments have the police powers to enforce public health measures; will there be police in hospitals to enforce triage decisions about ventilators? And how can the risk of violence be minimized?

How Should Patients in Respiratory Failure Be Managed if Ventilation Is Not Provided?

Patients with respiratory failure who do not receive mechanical ventilation during a pandemic are expected to die. They should receive respectful and compassionate palliative care. Dying from respiratory failure can be agonizing; patients commonly describe it as suffocating, drowning, or fighting for breath. Administering sedatives and analgesics is ethically and clinically appropriate in this situation (Lo and Rubenfeld, 2005). Even doses that will cause unconsciousness are appropriate if lower doses fail to relieve symptoms. Although such palliative sedation has strong ethical and legal justification, health-care workers are often confused about the distinction between palliative sedation, which is intended to relieve suffering, and active euthanasia, which is intended to kill the patient. Thus emergency-preparedness plans should include provisions for training physicians and nurses about palliative sedation and for providing emotional and spiritual support to patients, families, and health-care workers. Furthermore, shortages of resources besides ventilators might occur during a pandemic, so there may not be enough trained nurses to increase the dose of sedation and analgesia if lower doses have failed to relieve the suffering of dying patients, and disruptions to hospital supply chains may cause shortages of medications needed to relieve symptoms.

In summary, a number of general principles for protecting public health during a public health emergency have been articulated and, in some cases, enacted into state laws. Still, hospitals and frontline physicians need more specific criteria to triage patients with respiratory failure if a shortage of respirators develops. Furthermore, guidelines and procedures are needed to address the practical problems that will arise when putting triage priorities into practice. During a pandemic, it will not be feasible to carry out extensive discussions, so preparedness planning should anticipate the strong likelihood of a shortage of ventilators and develop explicit triage criteria and procedures ahead of time. Such discussions will need to involve the public in order to foster acceptance of the idea that during a pandemic some patients will die who might have been saved if they had received a ventilator.

Acknowledgments

This work was supported in part by the Greenwall Foundation and by NIH grant Roadmap K12 HD049077.

REFERENCES

Baker R, Strosberg M. 1992. Triage and equality: An historical reassessment of utilitarian analyses of triage. *Kennedy Institute of Ethics Journal* 2(2):103-123.

Bartlett JG. 2006. Planning for avian influenza. *Annals of Internal Medicine* 145(2):141-144.

Beauchamp TL, Childress JF. 2001. *Principles of Biomedical Ethics*. 5th ed. New York: Oxford University Press. Pp. 139-143, 165-176.

Bellagio Group. 2006a. *Bellagio Statement of Principles*. [Online]. Available: http://www.hopkinsmedicine.org/bioethics/bellagio/statement.html [accessed January 16, 2007].

Bellagio Group. 2006b. *Checklists for Avian and Pandemic Influenza Preparedness and Response*. [Online]. Available: http://www.hopkinsmedicine.org/bioethics/bellagio/checklist.html [accessed January 16, 2007].

Blendon RJ, Benson JM, Weldon KJ, Herrmann MJ. 2006. *Pandemic Influenza Survey*. [Online]. Available: http://www.hsph.harvard.edu/press/releases/press10262006.html [accessed January 16, 2007].

Burke TA, Shalauta NM, Tran NL, Stern BS. 1997. The environmental web: A national profile of the state infrastructure for environmental health and protection. *Journal Public Health Management and Practice* 3(2):1-12.

CDC (Centers for Disease Control and Prevention). 2003. *CDC News Conference Transcript: Influenza Update*. [Online]. Available: http://www.cdc.gov/od/oc/media/transcripts/t031219.htm [accessed February 7, 2007].

CDC. 2004. *CDC News Conference Transcript: Influenza Update*. [Online]. Available: http://www.cdc.gov/od/oc/media/transcripts/t040108.htm [accessed February 7, 2007].

CDC. 2006 (December 18). *CDC Letter to States from Dr. Richard Besser, Director, Coordinating Office of Terrorism Preparedness and Emergency Response*. [Online]. Available: http://www.hhs.gov/ophep/documents/Panfluopstemplate.pdf [accessed February 7, 2007].

CDC. 2007. *CDC FoodNet—Foodborne Diseases Active Surveillance Network Website*. [Online]. Available: http://www.cdc.gov/foodnet/#objectives [accessed February 6, 2007].

Collins SDW, Frost WH, Gover M, Sydenstricker E. 1930. Mortality from influenza and pneumonia in 50 large cities of the United States, 1910-1929. *Public Health Reports* 45(39):2277-2324.

Covello V, Sandman P. 2001. Risk communication: Evolution and revolution. In: Wolbarst A, ed. *Solutions to an Environment in Peril*. Baltimore, MD: Johns Hopkins University Press.

CSTE (Council on State and Territorial Epidemiologists). 2007a. *National Reporting for Novel Influenza A Virus Infections*. [Online]. Available: http://www.cste.org/PS/2007pdfs/NovelFluANNDSSJan10final23.pdf [accessed February 7, 2007].

CSTE. 2007b. *CSTE State Pandemic Influenza Plans*. [Online]. Available: http://www.cste.org/specialprojects/Influenzaplans/StateMap.asp [accessed February 6, 2007].

Cummings DA. In preparation. *A Review of the 1918 Pandemic in Developing and Developed Settings*. July 24, 2006.

Emanuel EJ, Wertheimer A. 2006. Public health: Who should get influenza vaccine when not all can? *Science* 312(5775):854-855.

Environmental Health Tracking Project Team. 2000. *America's Environmental Health Gap: Why the Country Needs a Nationwide Health Tracking Network*. Report sponsored by the Pew Environmental Health Commission, Johns Hopkins School of Public Health. [Online]. Available: http://healthyamericans.org/reports/files/healthgap.pdf [accessed February 7, 2007].

EPA (Environmental Protection Agency). 2007. *Safe Drinking Water Information System* (SDWIS). [Online]. Available: http://www.epa.gov/enviro/html/sdwis/sdwis_query.html [accessed January 23, 2007].

Erickson JD. 2000. Introduction: Birth defect surveillance in the United States. *Teratology* 61(1-2):1-3. [Online]. Available: http://www.nbdpn.org/archives/2000/report2000/1intro.pdf [accessed February 7, 2007].

Exploring Accreditation Project. 2006. *Final Recommendations for a Voluntary National Accreditation Program*. [Online]. Available: http://www.exploringaccreditation.org/doc/EAFinalRecommendations10-27-06.pdf [accessed February 7, 2007].

Ferguson NM, Cummings DA, Cauchemez S, Fraser C, Riley S, Meeyai A, Iamsirithaworn S, Burke DS. 2005. Strategies for containing an emerging influenza pandemic in Southeast Asia. *Nature* 437(7056):209-214.

Ferguson NM, Cummings DA, Fraser C, Cajka JC, Cooley PC, Burke DS. 2006. Strategies for mitigating an influenza pandemic. *Nature* 442(7101):448-452.

Gostin LO. 2000. *Public Health Law: Power, Duty, Restraint.* Berkeley: University of California Press.

Gostin LO. 2002. *Public Health Law and Ethics*. Berkeley and Los Angeles: University of California Press.

Gostin LO. 2004. Pandemic influenza: Public health preparedness for the next global health emergency. *Journal of Law, Medicine, and Ethics* 32(4):565-573.

Gostin LO. 2006. Medical countermeasures for pandemic influenza: Ethics and the law. *Journal of the American Medical Association* 295(5):554-556.

Graf J, Janssens U. 2005. The early bird catches the worm. *Critical Care Medicine* 33(10):2412-2414.

IOM (Institute of Medicine). 2002. *The Future of the Public's Health in the 21st Century*. Washington, DC: The National Academies Press.

Jajosky RA, Hall PA, Adams DA, Dawkins FJ, Sharp P, Anderson WJ, Aponte JJ, Jones GF, Nitschke DA, Worsham CA, Adekoya N, Doyle T. 2006. Summary of notifiable diseases—United States, 2004. *Morbidity and Mortality Weekly Report* 53(53):1-79. [Online]. Available: http://0-www.cdc.gov.mill1.sjlibrary.org/MMWR/PDF/wk/mm5353.pdf [accessed February 7, 2007].

Knaus WA. 2002. APACHE 1978-2001: The development of a quality assurance system based on prognosis: Milestones and personal reflections. *Archives of Surgery* 137(1):37-41.

Lasker RD. 2004. *Redefining Readiness: Terrorism Planning Through the Eyes of the Public.* New York: The New York Academy of Medicine. [Online]. Available: http://www.cacsh.org/pdf/RedefiningReadinessStudy.pdf [accessed February 7, 2007].

Lasker, RD. 2006 (December 6). *Keynote Address at New York Academy of Medicine Annual Meeting of Voting Fellows*. [Online]. Available: http://www.nyam.org/news/2825.html [accessed February 7, 2007].

Lo B. 2005. *Resolving Ethical Dilemmas: A Guide for Clinicians.* 3rd ed. Philadelphia, PA: Lippincott Williams & Wilkins: Pp. 28-35, 105-110.

Lo B, Katz MH. 2005. Clinical decision making during public health emergencies: Ethical considerations. *Annals of Internal Medicine* 143(7):493-498.

Lo B, Rubenfeld G. 2005. Palliative sedation in dying patients: "We turn to it when everything else hasn't worked." *Journal of the American Medical Association* 294(14):1810-1816.

Longini IM, Jr., Nizam A, Xu S, Ungchusak K, Hanshaoworakul W, Cummings DA, Halloran ME. 2005. Containing pandemic influenza at the source. *Science* 309(5737):1083-1087.

Lurie N, Wasserman J, Stoto M, Myers S, Namkung P, Fielding J, Valdez RB. 2004. Local variation in public health preparedness: Lessons from California. *Health Affairs* Jan-Jun(Suppl Web Exclusive):W4-341-353.

Mills ID. 1986. The 1918-1919 influenza pandemic—The Indian experience. *Indian Economic and Social History Review* 23(1):1-40.

Murray CJ, Lopez AD, Chin B, Feehan D, Hill KH. 2006. Estimation of potential global pandemic influenza mortality on the basis of vital registry data from the 1918-20 pandemic: A quantitative analysis. *Lancet* 368(9554):2211-2218.

Phillips H. 1990. *Black October: The Impact of the Spanish Influenza Epidemic of 1918 on South Africa*. Pretoria, South Africa: CTP Book Printers.

Powers M, Faden R. 2006. *Social Justice: The Moral Foundations of Public Health and Health Policy*. New York: Oxford University Press.

Registrar General of the United Kingdom. 1920. *Report of the Mortality From Influenza in England and Wales During the Epidemic of 1918-1919: Supplement to the Eighty-First Annual Report of the Registrar General of Births, Deaths, and Marriages in England and Wales*. London, UK: Her Majesty's Stationery Office.

Repine TB, Lisagor P, Cohen DJ. 2005. The dynamics and ethics of triage: Rationing care in hard times. *Military Medicine* 170(6):505-509.

Society of Critical Care Medicine Ethics Committee. 1994. Consensus statement on the triage of critically ill patients. *Journal of the American Medical Association* 271(15):1200-1203.

Sutton V. 2001. Bioterrorism preparation and response legislation—The struggle to protect states' sovereignty while preserving national security. *The Georgetown Public Policy Review* 6:93.

Sutton V. 2005. A multidisciplinary approach to an ethic of biodefense and bioterrorism. *Journal of Law, Medicine, and Ethics* 33(2):310-322.

Sydenstricker E. 1931. The incidence of influenza among persons of different economic status during the epidemic of 1918. *Public Health Reports* 46(4):154-170.

Theilen L. 2004. *Accreditation of Public Health Agencies*. Robert Wood Johnson Foundation report.

Trust for America's Health. 2005. *Birth Defects and Developmental Disabilities: The Search for Causes and Cures*. Washington, DC: Trust for America's Health. [Online]. Available: http://healthyamericans.org/reports/birthdefects05/BirthDefects05.pdf [accessed February 7, 2007].

Trust for America's Health. 2006. *Ready or Not? Protecting the Public's Health from Diseases, Disasters and Bioterrorism*. Washington, DC: Trust for American's Health. [Online]. Available: http://healthyamericans.org/reports/bioterror06/BioTerrorReport2006.pdf [accessed February 7, 2007].

Uhlman, M. 2005 (May 19). PA has a plan—but it's top secret. *Philadelphia Inquirer*.

UNAIDS (Joint United Nations Programme on HIV/AIDS) and World Health Organization. 2004. *Guidance on Ethics and Equitable Access to HIV Treatment and Care*. WHO: Geneva. [Online]. Available: http://www.who.int/ethics/en/ethics_equity_HIV_e.pdf [accessed April 6, 2007].

Veatch RM. 2005. Disaster preparedness and triage: Justice and the common good. *Mt. Sinai Journal of Medicine* 72(4):236-241.

Appendix A

Agenda

Ethical and Legal Considerations in Mitigating Pandemic Disease
PAHO Headquarters—Conference Room A
525 23rd Street, NW
Washington, D.C.
September 19-20, 2006

AGENDA

Tuesday, September 19, 2006

8:30 a.m.	Continental Breakfast
9:00 a.m.	Welcome and Opening Remarks Stanley M. Lemon, M.D., Chair, Forum on Microbial Threats
9:15–10:00 a.m.	KEYNOTE ADDRESS: "Past as Prologue?" David Heymann, M.D., World Health Organization
10:00–10:15 a.m.	Discussion

Session I:
Understanding the Challenges of the Future by Examining the Past: Influenza/Smallpox/SARS

Moderator: Ruth Berkelman, M.D., Emory University, Rollins School of Public Health

10:15–11:00 a.m.	Contemplating Pandemics: The Role of Historical Inquiry in Developing Pandemic-Mitigation Strategies for the Twenty-First Century Howard Markel, M.D., University of Michigan
11:00–11:45 a.m.	The Smallpox Eradication Campaign D.A. Henderson, M.D., Center for Biosecurity, University of Pittsburgh Medical Center
11:45 a.m.–12:30 p.m.	Lessons from SARS David Heymann, M.D., World Health Organization
12:30–1:00 p.m.	Session I Q&A
1:00–1:45 p.m.	LUNCH

Session II:
Domestic, Regional, and International Preparedness Planning

Moderator: P. Frederick Sparling, M.D., Vice Chair, Forum on Microbial Threats

1:45–2:30 p.m.	U.S. Government Preparedness Plans Bruce Gellin, M.D., Office of the Secretary, Department of Health and Human Services
2:30–3:15 p.m.	Regional Planning Efforts Oscar J. Mujica, M.D., M.P.H., P.H.E., Pan American Health Organization
3:15–3:45 p.m.	Q&A
3:45–4:00 p.m.	BREAK
4:00–4:45 p.m.	Ethical Considerations in International Preparedness Planning Efforts Alexander Morgan Capron, University of Southern California

4:45–5:15 p.m.	Open Discussion of Session II
5:15–6:00 p.m.	Open Discussion of Day 1
6:00 p.m.	Adjourn

Wednesday, September 20, 2006

8:00 a.m.	Continental Breakfast
8:30 a.m.	Opening Remarks/Summary of Day 1: Margaret A. Hamburg, M.D., Vice Chair, Forum on Microbial Threats

Session III:
Disease Mitigation Strategies–Quarantine, Containment and Modeling

Moderator: Gary Roselle, M.D., Program Director for Infectious Diseases, Department of Veterans Affairs Central Office

8:45–9:30 a.m.	Public Health and Ethical Considerations in Planning for Quarantine Martin Cetron, M.D., Centers for Disease Control and Prevention
9:30–10:15 a.m.	Preparing for Pandemic Influenza: Legal and Ethical Challenges Lawrence Gostin, J.D., Georgetown University School of Law
10:15–10:30 a.m.	Discussion
10:30–11:15 a.m.	The Role of Modeling in Infectious Disease Mitigation and Containment Joshua Epstein, Ph.D., The Brookings Institution
11:15–11:45 a.m.	Discussion Panel: • Timothy C. Germann, Ph.D., Los Alamos National Laboratory • James LeDuc, M.D., Centers for Disease Control and Prevention • Victoria Sutton, J.D., Ph.D., Texas Tech University School of Law

11:45 a.m.–12:30 p.m.	Lunch
12:30–1:30 p.m.	Luncheon Remarks Harvey V. Fineberg, M.D., Ph.D., President, Institute of Medicine

Session IV:
Ethical Issues in Pandemic Planning and Response

Moderator: Stanley M. Lemon, M.D., Chair, Forum on Microbial Threats

1:30–2:15 p.m.	Intensive Care Unit Triage During an Influenza Pandemic: The Need for Specific Clinical Guidelines Bernard Lo, M.D., University of California, San Francisco
2:15–3:00 p.m.	Social Justice and Pandemic Planning and Response Ruth Faden, Ph.D., Berman Bioethics Institute, Johns Hopkins University
3:00–3:45 p.m.	Discussion Panel: • Steven Bice, Battelle, Atlanta, Georgia • D.A. Henderson, M.D., Center for Biosecurity, University of Pittsburgh Medical Center • Shelley Hearne, D.P.H., Johns Hopkins University
3:45–4:30 p.m.	Open Discussion
4:30–4:45 p.m.	Closing Remarks/Adjourn

Appendix B

Acronyms

ABM	agent-based models
ACIP	Advisory Committee on Immunization Practices
AIDS	Acquired Immunodeficiency Syndrome
APEC	Asia-Pacific Economic Cooperation
APHA	American Public Health Association
ARV	Antiretroviral
ASEAN	Association of Southeast Asian Nations
BSE	bovine spongiform encephalopathy
CDC	Centers for Disease Control and Prevention
CISAC	Committee on International Security and Arms Control
CSTE	Council of State and Territorial Epidemiologists
DHS	U.S. Department of Homeland Security
EPA	U.S. Environmental Protection Agency
FAO	Food and Agriculture Organization
GEIS	Global Emerging Infections System
GOARN	Global Outbreak Alert and Response Network
GPHIN	Global Public Health Intelligence Network

HHS	Department of Health and Human Services
HIV	Human Immunodeficiency Virus
HRSA	Health Resources and Services Administration
HSC	Homeland Security Council
ICU	Intensive Care Unit
IHR	International Health Regulations
IOM	Institute of Medicine
MIDAS	Models of Infectious Disease Agent Study
NGO	Nongovernmental organization
NIH	National Institutes of Health
NIPPP	National Influenza Pandemic Preparedness Plans
NPI	Nonpharmaceutical Interventions
NSPI	National Strategy for Pandemic Influenza
NVAC	National Vaccine Advisory Committee
NYAM	New York Academy of Medicine
OIE	Office International des Epizooties (World Organization for Animal Health)
PAHO	Pan-American Health Organization
PATH	Program for Appropriate Technology in Health
PPE	personal protective equipment
SARS	Severe Acute Respiratory Syndrome
SNS	Strategic National Stockpile
UN	United Nations
UNAIDS	United Nations Programme on AIDS
UNESCO	United Nations Educational, Scientific, and Cultural Organization
UNICEF	United Nations Children's Fund
USAID	U.S. Agency for International Development
WHO	World Health Organization

Appendix C

Forum Member Biographies

Stanley M. Lemon, M.D. *(Chair),* is the John Sealy Distinguished University Chair and director of the Institute for Human Infections and Immunity at the University of Texas Medical Branch (UTMB) at Galveston. He received his undergraduate A.B. degree in biochemical sciences from Princeton University summa cum laude and his M.D. with honor from the University of Rochester. He completed postgraduate training in internal medicine and infectious diseases at the University of North Carolina at Chapel Hill and is board certified in both. From 1977 to 1983 he served with the U.S. Army Medical Research and Development Command, followed by a 14-year period on the faculty of the University of North Carolina School of Medicine. He moved to UTMB in 1997, serving first as chair of the Department of Microbiology and Immunology, then as dean of the School of Medicine from 1999 to 2004. Dr. Lemon's research interests relate to the molecular virology and pathogenesis of the positive-stranded RNA viruses responsible for hepatitis. He has had a long-standing interest in antiviral and vaccine development and has served previously as chair of the Anti-Infective Drugs Advisory Committee of the U.S. Food and Drug Administration (FDA). He is the past chair of the Steering Committee on Hepatitis and Poliomyelitis of the World Health Organization (WHO) Programme on Vaccine Development. He presently serves as a member of the U.S. Delegation of the U.S.–Japan Cooperative Medical Sciences Program, and he chairs the Board of Scientific Councilors of the National Center for Infectious Diseases (NCID) of the Centers for Disease Control and Prevention (CDC). He was co-chair of the Committee on Advances in Technology and the Prevention of their Application to Next Generation Biowarfare Threats for the National Academy of Sciences (NAS) and recently chaired an Institute of Medicine (IOM) study committee related to vaccines for the protection of the military against naturally occurring infectious disease threats.

Margaret A. Hamburg, M.D. (*Vice-chair*), is vice president for Biological Programs at the Nuclear Threat Initiative, a charitable organization working to reduce the global threat from nuclear, biological, and chemical weapons. She is in charge of the biological program area. She completed her internship and residency in internal medicine at the New York Hospital/Cornell University Medical Center and is certified by the American Board of Internal Medicine. Dr. Hamburg is a graduate of Harvard College and Harvard Medical School. Before taking on her current position, she was the assistant secretary for planning and evaluation, U.S. Department of Health and Human Services (HHS), serving as a principal policy advisor to the secretary of health and human services with responsibilities including policy formulation and analysis, the development and review of regulations and legislation, budget analysis, strategic planning, and the conduct and coordination of policy research and program evaluation. Prior to this, she served for almost six years as the commissioner of health for the city of New York. As chief health officer in the nation's largest city, her many accomplishments included the design and implementation of an internationally recognized tuberculosis control program that produced dramatic declines in tuberculosis cases, the development of initiatives that raised childhood immunization rates to record levels, and the creation of the first public health bioterrorism preparedness program in the nation. She currently serves on the Harvard University Board of Overseers. She has been elected to membership in the IOM, the New York Academy of Medicine, and the Council on Foreign Relations and is a fellow of the American Association for the Advancement of Science (AAAS) and the American College of Physicians.

P. Frederick Sparling, M.D. (*Vice-chair*), is the J. Herbert Bate Professor Emeritus of Medicine, Microbiology, and Immunology at the University of North Carolina (UNC) at Chapel Hill and is director of the North Carolina Sexually Transmitted Infections Research Center. Previously he served as chair of the Department of Medicine and chair of the Department of Microbiology and Immunology at UNC. He was president of the Infectious Disease Society of America from 1996 to 1997. He was also a member of the IOM's Committee on Microbial Threats to Health (1991–1992). Dr. Sparling's laboratory research is in the molecular biology of bacterial outer membrane proteins involved in pathogenesis, with a major emphasis on *gonococci* and *meningococci*. His current studies focus on the biochemistry and genetics of iron-scavenging mechanisms used by *gonococci* and *meningococci* and the structure and function of the *gonococcal porin* proteins. He is pursuing the goal of a vaccine for gonorrhea.

David W. K. Acheson, M.D., is chief medical officer at the FDA's Center for Food Safety and Applied Nutrition. He received his medical degree at the University of London. After completing internships in general surgery and medicine, he continued his postdoctoral training in Manchester, England, as a Wellcome Trust research fellow. He subsequently was a Wellcome Trust training fellow in Infectious Diseases at the New England Medical Center and at the Wellcome Research Unit in Vellore, India. He was associate professor of medicine, Division of Geographic Medicine and Infectious Diseases, New England Medical Center, until 2001. He then joined the faculties of the Department of Epidemiology and Preventive Medicine and Department of Microbiology and Immunology at the University of Maryland Medical School. Currently at the FDA, Dr. Acheson's research concentration is on foodborne pathogens and encompasses a mixture of molecular pathogenesis, cell biology, and epidemiology. Specifically, his research focuses on Shiga toxin-producing *E. coli* and understanding toxin interaction with intestinal epithelial cells using tissue culture models. His laboratory has also undertaken a study to examine Shiga toxin-producing *E. coli* in food animals in relation to virulence factors and antimicrobial resistance patterns. More recently, Dr. Acheson initiated a project to understand the molecular pathogenesis of *Campylobacter jejuni*. Other studies have undertaken surveillance of diarrheal disease in the community to determine causes, outcomes, and risk factors of unexplained diarrhea. Dr. Acheson has authored or coauthored more than 72 journal articles and 42 book chapters and reviews, and he is coauthor of the book *Safe Eating* (Dell Health, 1998). He serves as a reviewer for more than 10 journals and is on the editorial boards of *Infection and Immunity* and *Clinical Infectious Diseases*. He is a fellow of the Royal College of Physicians and a fellow of the Infectious Disease Society of America, and he holds several patents.

Ruth L. Berkelman, M.D., is the Rollins Professor and director of the Center for Public Health Preparedness and Research at the Rollins School of Public Health, Emory University in Atlanta. She received her A.B. from Princeton University and her M.D. from Harvard Medical School. Board certified in pediatrics and internal medicine, she began her career at the CDC in 1980 and later became deputy director of the NCID. She also served as a senior advisor to the director, CDC, and as Assistant Surgeon General in the U.S. Public Health Service. In 2001 she came to her current position at Emory University, directing a center focused on emerging infectious disease and other urgent threats to health, including terrorism. She has also consulted with the biologic program of the Nuclear Threat Initiative and is most recognized for her work in infectious diseases and disease surveillance. She was elected to the IOM in 2004. Currently a member of the Board on Life Sciences of the National Academies, she also chairs the Board of Public and Scientific Affairs at the American Society of Microbiology.

Enriqueta C. Bond, Ph.D., is president of the Burroughs Wellcome Fund. She received her undergraduate degree from Wellesley College, her M.A. from the University of Virginia, and her Ph.D. in molecular biology and biochemical genetics from Georgetown University. She is a member of the IOM, the AAAS, the American Society for Microbiology (ASM), and the American Public Health Association. Dr. Bond serves on the council of the IOM as its vice-chair; she chairs the Board of Scientific Counselors for the NCID at the CDC, and she chairs the IOM's Clinical Research Roundtable. She serves on the board and the executive committee of the Research Triangle Park Foundation and on the board of the Medicines for Malaria Venture. Prior to being named president of the Burroughs Wellcome Fund in 1994, she had served on the staff of the IOM since 1979, becoming the IOM's executive officer in 1989.

Roger G. Breeze, Ph.D., received his veterinary degree in 1968 and his Ph.D. in veterinary pathology in 1973, both from the University of Glasgow, Scotland. He was engaged in teaching, diagnostic pathology, and research on respiratory and cardiovascular diseases at the University of Glasgow Veterinary School from 1968 to 1977 and at Washington State University College of Veterinary Medicine from 1977 to 1987, where he was professor and chair of the Department of Microbiology and Pathology. From 1984 to 1987 he was deputy director of the Washington Technology Center, the state's high-technology sciences initiative, based in the College of Engineering at the University of Washington. In 1987, he was appointed director of the USDA's Plum Island Animal Disease Center, a biosafety level 3 facility for research and diagnosis of the world's most dangerous livestock diseases. In that role he initiated research into the genomic and functional genomic basis of disease pathogenesis, diagnosis, and control of livestock RNA and DNA virus infections. This work became the basis of U.S. defense against natural and deliberate infection with these agents and led to his involvement in the early 1990s in biological weapons defense and proliferation prevention. From 1995 to 1998 he directed research programs in 20 laboratories in the Southeast for the USDA Agricultural Research Service before going to Washington, DC, to establish biological weapons defense research programs for the USDA. He received the Distinguished Executive Award from President Clinton in 1998 for his work at Plum Island and in biodefense. Since 2004 he has been CEO of Centaur Science Group, which provides consulting services in biodefense. His main commitment is to the Defense Threat Reduction Agency's Biological Weapons Proliferation Prevention program in Europe, the Caucasus, and Central Asia.

Steven J. Brickner, Ph.D., is research advisor, antibacterials chemistry, at Pfizer Global Research and Development. He received his Ph.D. in organic chemistry from Cornell University and was a National Institutes of Health (NIH) postdoctoral research fellow at the University of Wisconsin-Madison. He is a medicinal chemist with nearly 20 years of research experience in the pharmaceutical industry, all focused on the discovery and development of novel antibacterial agents. He is an inventor or coinventor on 21 U.S. patents and has published numerous scientific papers, primarily within the area of the oxazolidinones. Prior to joining Pfizer in 1996, he led a team at Pharmacia and Upjohn that discovered and developed linezolid, the first member of a new class of antibiotics to be approved in the last 35 years.

Nancy Carter-Foster, M.S.T.M., is senior advisor for health affairs for the U.S. Department of State, assistant secretary for science and health, and the secretary's representative on HIV/AIDS. She is responsible for identifying emerging health issues and making policy recommendations for the United States foreign policy concerns regarding international health, and she coordinates the department's interactions with the nongovernmental community. She is a member of the Infectious Diseases Society of America (IDSA) and the AAAS. She has helped bring focus to global health issues in U.S. foreign policy and has brought a national security focus to global health. In prior positions as director for congressional and legislative affairs for the Economic and Business Affairs Bureau of the U.S. Department of State, foreign policy advisory to the majority whip of the U.S. House of Representatives, trade specialist advisor to the House of Representatives Ways and Means Trade Subcommittee, and consultant to the World Bank, Asia Technical Environment Division, Ms. Carter-Foster has worked on a wide variety of health, trade, and environmental issues amassing in-depth knowledge and experience in policy development and program implementation.

Gail H. Cassell, Ph.D., is vice president of Scientific Affairs, Distinguished Lilly Research Scholar for Infectious Diseases, Eli Lilly & Company. Previously she was the Charles H. McCauley Professor and, beginning in 1987, the chair of the Department of Microbiology, University of Alabama Schools of Medicine and Dentistry at Birmingham, a department which, under her leadership, ranked first in research funding from the NIH since 1989. She is a member of the Director's Advisory Committee of the CDC. Dr. Cassell is past president of the ASM and is serving her third 3-year term as chair of the Public and Scientific Affairs Board of the ASM. She is a former member of the NIH Director's Advisory Committee and a former member of the Advisory Council of the National Institute of Allergy and Infectious Diseases. She has also served as an advisor on infectious diseases and indirect costs of research to the White House Office on Science and Technology and was previously chair of the Board of Scientific Counselors of the NCID at the CDC. She served eight years on the Bacteriology-Mycology-II Study Section and served as its chair for three years. She serves on the editorial boards of several prestigious scientific journals and has authored over 275 articles and book chapters. She has been intimately involved in the establishment of science policy and legislation related to biomedical research and public health. Dr. Cassell has received several national and international awards and an honorary degree for her research on infectious diseases.

Bill Colston, Ph.D., is currently the Division Leader for the Chemical and Biological Countermeasures (CB) Division at Lawrence Livermore National Laboratory (LLNL). This newly formed division consists of four programs whose mission's include threat awareness, detection, response, and attribution. These programs are made up of ~190 researchers from a variety of disciplines. The mission of these programs is to provide science, technology, and deployed capabilities to defend our nation, its people and warfighters, against the threat of biological and chemical terrorism. The larger vision is to meet the challenges of an ever-changing threat by transforming our understanding of pathogenicity and host response; and, expanding our reach globally. Dr. Colston holds a Ph.D in Biomedical Engineering and has published numerous publications and patents, largely in biological measurement sciences. Directly prior to this assignment he founded the Department of Homeland Security's Biodefense Knowledge Center (BKC).

Col. Ralph (Loren) Erickson, M.D., Dr.P.H., M.P.H., is the director of the Department of Defense Global Emerging Infections Surveillance and Response System (DoD-GEIS) headquartered in Silver Spring, Maryland. He holds a B.S. degree in chemistry from the University of Washington, an M.D. from the Uniformed Services University of the Health Sciences, an M.P.H. from Harvard, and a Dr.P.H. from Johns Hopkins. Residency trained and board certified in preventive medicine, Dr. Erickson has held a number of leadership positions within the Army Medical Department, including: director of the General Preventive Medicine Residency Program, Walter Reed Army Institute of Research; director of Epidemiology and Disease Surveillance, U.S. Army Center for Health Promotion and Preventive Medicine; commander of the U.S. Army Center for Health Promotion and Preventive Medicine (Europe); and specialty leader for all U.S. Army preventive medicine physicians.

Mark Feinberg, M.D., Ph.D., is vice president for Policy, Public Health, and Medical Affairs in the Merck Vaccine Division of Merck & Co., Inc. He received his bachelor's degree magna cum laude from the University of Pennsylvania in 1978 and his M.D. and Ph.D. degrees from Stanford University School of Medicine in 1987. From 1985 to 1986, Dr. Feinberg served as a project officer for the Committee on a National Strategy for AIDS of the IOM and the NAS. Following receipt of his M.D. and Ph.D. degrees, he pursued postgraduate residency training in internal medicine at the Brigham and Women's Hospital of Harvard Medical School and postdoctoral fellowship research in the laboratory of Dr. David Baltimore at the Whitehead Institute for Biomedical Research. From 1991 to 1995, Dr. Feinberg was an assistant professor of medical and microbiology and immunology at the University of California, San Francisco (UCSF), where he also served as an attending physician in the AIDS/Oncology Division and as director of the Virology Research Laboratory at San Francisco General Hospital. From 1995 to

1997, he was a medical officer in the Office of AIDS Research in the office of the director of the NIH, and chair of the NIH Coordinating Committee on AIDS Etiology and Pathogenesis Research. During this period, he also served as executive secretary of the NIH Panel to Define Principles of Therapy of HIV Infection. Prior to joining Merck in 2004, Dr. Feinberg served as professor of medicine and microbiology and immunology at the Emory University School of Medicine and as an investigator at the Emory Vaccine Center. He also founded and served as the medical director of the Hope Clinic—a clinical research facility devoted to the clinical evaluation of novel vaccines and to translational research studies of human immune system biology. At UCSF and Emory, Dr. Feinberg and colleagues were engaged in the preclinical development and evaluation of novel vaccines for HIV and other infectious diseases and in basic research studies focused on revealing fundamental aspects of host-virus relationships that underlie the pathogenesis of HIV and simian immunodeficiency virus (SIV) infections. In addition to his other professional roles, he has also served as a consultant to, and member of, several committees of the IOM and the NAS.

J. Patrick Fitch, Ph.D., is Laboratory Director for the National Biodefense Analysis and Countermeasures Center (NBACC) and the President of Battelle National Biodefense Institute, LLC (BNBI). BNBI manages and operates the NBACC national laboratory for the Department of Homeland Security as a Federally Funded Research and Development Center established in 2006. The NBACC mission is to provide the nation with the scientific basis for awareness of biological threats and attribution of their use against the American public. Dr. Fitch joined Battelle in 2006 as vice president for Biodefense Programs after more than 20 years of experience leading multidisciplinary applied-science teams at the University of California's Lawrence Livermore National Laboratory (LLNL). From 2001 to 2006, he led the LLNL Chemical and Biological National Security Program (CBNP), with applied science programs from pathogen biology and material science to deployed systems. CBNP accomplishments include performing more than one million assays on national security samples; set up and operation of 24/7 reach-back capabilities; set up of a nationwide bio-alert system; three R&D 100 awards; design of signatures for validated assays in the CDC Laboratory Response Network and the National Animal Health Laboratory Network; and the BASIS biodetection system was designed, demonstrated, and deployed, leading to the nationwide BioWatch system. He has authored several books and book chapters, including *An Engineering Introduction to Biotechnology*. He has chaired and served on several panels of the National Academies. His advisory board activities have included U.S. Animal Health Association, Texas A&M University DHS Center of Excellence, Central Florida University (College Engineering), Colorado State University (College of Engineering), California State Breast Cancer Research Program, and *Biomolecular Engineering*. Dr. Fitch was a Fellow of the American Society for Laser Medicine and

Surgery and an Associate Editor of *Circuits, Systems and Signal Processing*. He has received two national awards for medical devices, a technical writing award for an article in *Science*, and an international best paper award from the IEEE. He also co-invented the technology, developed the initial business plan, and successfully raised venture investments for a high tech medical device start-up company. Dr. Fitch received his Ph.D. from Purdue University and B.S. degrees from Loyola College of Maryland.

Capt. Darrell R. Galloway, MSC, Ph.D., is Chief of Medical S&T Division for the Chemical & Biological Defense Directorate at the Defense Threat Reduction Agency. He received his baccalaureate degree in microbiology from California State University in Los Angeles in 1973. After completing military service in the U.S. Army as a medical corpsman from 1969 to 1972, Captain Galloway entered graduate school and completed a doctoral degree in biochemistry in 1978 from the University of California, followed by two years of postgraduate training in immunochemistry as a Fellow of the National Cancer Institute at the Scripps Clinic and Research Foundation in La Jolla, California. Captain Galloway began his Navy career at the Naval Medical Research Institute in Bethesda, Maryland, where from 1980 to 1984 he served as a research scientist working on vaccine development. In late 1984 Captain Galloway left active service to pursue an academic appointment at the Ohio State University, where he is now a tenured faculty member in the Department of Microbiology. He also holds appointments at the University of Maryland Biotechnology Institute and the Uniformed Services University of Health Sciences. He has an international reputation in the area of bacterial toxin research and has published more than 50 research papers on various studies of bacterial toxins. In recent years Captain Galloway's research has concentrated on anthrax and the development of DNA-based vaccine technology. His laboratory has contributed substantially to the development of a new DNA-based vaccine against anthrax which has completed the first phase of clinical trials. Captain Galloway is a member of the ASM and has served as president of the Ohio branch of that organization. He received an NIH Research Career Development Award. In 2005 Captain Galloway was awarded the Joel M. Dalrymple Award for significant contributions to biodefense vaccine development.

S. Elizabeth George, Ph.D., is deputy director, Biological Countermeasures Portfolio Science and Technology Directorate, Department of Homeland Security (DHS). Until merging into the new department in 2003, she was the program manager of the Chemical and Biological National Security Program in the Department of Energy's National Nuclear Security Administration's Office of Nonproliferation Research and Engineering. Significant accomplishments include the design and deployment of BioWatch, the nation's first civilian biological-threat-agent-monitoring system and PROTECT, the first civilian operational chemical detection and response capability deployed in the Washington, D.C.-area subway system.

Previously, she spent 16 years at the U.S. Environmental Protection Agency (EPA), Office of Research and Development, National Health and Ecological Effects Research Laboratory, Environmental Carcinogenesis Division, where she was branch chief of the Molecular and Cellular Toxicology Branch. She received her B.S. in biology in 1977 from Virginia Polytechnic Institute and State University and her M.S. and Ph.D. in microbiology in1979 and 1984, respectively, from North Carolina State University. From 1984 to 1986 she was a National Research Council fellow in the laboratory of Dr. Larry Claxton at the EPA. Dr. George is the 2005 chair of the Chemical and Biological Terrorism Defense Gordon Research Conference. She has served as councilor for the Environmental Mutagen Society and president and secretary of the Genotoxicity and Environmental Mutagen Society. She holds memberships in the ASM and the AAAS and is an adjunct faculty member in the School of Rural Public Health, Texas A&M University. She is a recipient of the EPA Bronze Medal and Scientific and Technological Achievement Awards and DHS Under Secretary's Award for Science and Technology. She is author of numerous journal articles and has presented her research at national and international meetings.

Jesse L. Goodman, M.D., M.P.H., is director of FDA's Center for Biologics Evaluation and Research (CBER) which oversees medical, public health, and policy activities concerning the development and assessment of vaccines, blood products, tissues, and related devices and novel therapeutics including cellular and gene therapies. He moved full-time to FDA in 2001 from the University of Minnesota, where he was professor of medicine and director of the Division of Infectious Diseases. A graduate of Harvard College, he received his M.D. at the Albert Einstein College of Medicine, did residency and fellowship training at the Hospital of the University of Pennsylvania and at UCLA (where he was also chief medical resident), and is board certified in internal medicine, oncology, and infectious diseases. He trained in the virology laboratory of Jack Stevens at UCLA and has had an active laboratory program in the molecular pathogenesis of infectious diseases. In 1995 his laboratory isolated the etiologic agent of human granulocytic ehrlichiosis (HGE) and subsequently characterized fundamental events involved in infection of leukocytes, including their cellular receptors. He is editor of the book "Tick Borne Diseases of Humans" published by ASM Press in 2005 and is a staff physician and infectious diseases consultant at the NIH Clinical Center and the National Naval Medical Center/Walter Reed Army Medical Center, as well as adjunct professor of medicine at the University of Minnesota. He is active in a wide variety of clinical, public health, and product-development issues, including pandemic and emerging infectious disease threats, bioterrorism preparedness and response, and blood, tissue and vaccine safety and availability. In these activities, he has worked closely with CDC, NIH, and other HHS components, academia and the private sector, and he has put into place an interactive team approach to emerging threats. This model was used in the collaborative development and rapid implementation of nationwide donor screening of the U.S. blood supply for West Nile virus. He has been elected to the American Society for Clinical Investigation (ASCI) and to the IOM.

Eduardo Gotuzzo, M.D., is principal professor and director at the Instituto de Medicina Tropical "Alexander von Humbolt," Universidad Peruana Cayetan Heredia (UPCH) in Lima, Peru, as well as chief of the Department of Infectious and Tropical Diseases at the Cayetano Heredia Hospital. He is also an adjunct professor of medicine at the University of Alabama, Birmingham School of Medicine. Dr. Gotuzzo is an active member in numerous international societies and has been president of the Latin America Society of Tropical Disease (2000-2003), the Scientific Program of Infectious Diseases Society of America (2000-2003), the International Organizing Committee of the International Congress of Infectious Diseases (1994-present), president elect of the International Society for Infectious Diseases (1996-1998), and president of the Peruvian Society of Internal Medicine (1991-1992). He has published over 230 articles and chapters as well as six manuals and one book. Recent honors and awards include being named an honorary member of American Society of Tropical Medicine and Hygiene in 2002, associate member of the National Academy of Medicine in 2002, honorary member of the Society of Internal Medicine in 2000, and distinguished visitor at the Faculty of Medical Sciences, University of Cordoba, Argentina in 1999, and in 1988 he received the Golden Medal for Outstanding Contribution in the Field of Infectious Diseases awarded by the Trnava University, Slovakia.

Jo Handelsman, Ph.D., received her Ph.D. in molecular biology from the University of Wisconsin-Madison (UW-M) in 1984 and joined the faculty of the UW-M Department of Plant Pathology in 1985, where she is currently a Howard Hughes Medical Institute professor. Her research focuses on the genetic and functional diversity of microorganisms in soil and insect gut communities. The Handelsman lab has concentrated on discovery and biological activity of novel antibiotics from cultured and uncultured bacteria and has contributed to the pioneering of a new technique, called metagenomics, that facilitates the genomic analysis of assemblages of uncultured microorganisms. Handelsman is studying the midgut of the gypsy moth to understand the basis for resistance and susceptibility of microbial communities to invasion, developing it as a model for the microbial community in the human gut. In addition to her passion for understanding the secret lives of bacteria, Dr. Handelsman is dedicated to improving science education and the advancement of women in research universities. She is director of the Howard Hughes Medical Institute New Generation Program for Scientific Teaching, which is dedicated to teaching graduate students and postdoctoral students the principles and practices of teaching and mentoring. She is co-director of the National Academies Summer Institute for Undergraduate Education in Biology, which is a collaborative venture between HHMI and the National Academies that aims to train a nationwide network of faculty who are outstanding teachers and mentors. Dr. Handelsman is co-director of the Women in Science and Engineering Leadership Institute at UW-M, whose mission is to understand the impediments to the successful recruitment and advancement of women faculty in the sciences and to develop and study interventions intended to reduce the barriers.

Carole A. Heilman, Ph.D., is director of the Division of Microbiology and Infectious Diseases (DMID) of the National Institute of Allergy and Infectious Diseases (NIAID). She received her bachelor's degree in biology from Boston University in 1972 and earned her master's degree and doctorate in microbiology from Rutgers University in 1976 and 1979, respectively. Dr. Heilman began her career at the NIH as a postdoctoral research associate with the National Cancer Institute where she carried out research on the regulation of gene expression during cancer development. In 1986, she came to NIAID as the influenza and viral respiratory diseases program officer in DMID and, in 1988, she was appointed chief of the respiratory diseases branch where she coordinated the development of acellular pertussis vaccines. She joined the Division of AIDS as deputy director in 1997 and was responsible for developing the Innovation Grant Program for Approaches in HIV Vaccine Research. She is the recipient of several notable awards for outstanding achievement. Throughout her extramural career, Dr. Heilman has contributed articles on vaccine design and development to many scientific journals and has served as a consultant to the World Bank and WHO in this area. She is also a member of several professional societies, including the Infectious Diseases Society of America, the ASM, and the American Society of Virology.

David L. Heymann, M.D., is currently the executive director of the WHO Communicable Diseases Cluster. From October 1995 to July 1998 he was director of the WHO Programme on Emerging and Other Communicable Diseases Surveillance and Control. Prior to becoming director of this program, he was the chief of research activities in the Global Programme on AIDS. From 1976 to 1989, prior to joining WHO, Dr. Heymann spent 13 years working as a medical epidemiologist in sub-Saharan Africa (Cameroon, Ivory Coast, the former Zaire, and Malawi) on assignment from the CDC in CDC-supported activities aimed at strengthening capacity in surveillance of infectious diseases and their control, with special emphasis on the childhood immunizable diseases, African hemorrhagic fevers, pox viruses, and malaria. While based in Africa, he participated in the investigation of the first outbreak of Ebola in Yambuku in the former Zaire in 1976, then investigated the second outbreak of Ebola in 1977 in Tandala, and in 1995 directed the international response to the Ebola outbreak in Kikwit. Prior to 1976, Dr. Heymann spent two years in India as a medical officer in the WHO Smallpox Eradication Programme. He holds a B.A. from the Pennsylvania State University, an M.D. from Wake Forest University, and a Diploma in Tropical Medicine and Hygiene from the London School of Hygiene and Tropical Medicine. He has also completed practical epidemiology training in the Epidemic Intelligence Service (EIS) training program of the CDC. He has published 131 scientific articles on infectious diseases in peer-reviewed medical and scientific journals.

Phil Hosbach is vice president of New Products and Immunization Policy at Sanofi Pasteur. The departments under his supervision are new product marketing, state and federal government policy, business intelligence, bids and contracts, medical communications, public health sales, and public health marketing. His current responsibilities include oversight of immunization policy development. He acts as Sanofi Pasteur's principle liaison with CDC. Mr. Hosbach graduated from Lafayette College in 1984 with a degree in biology. He has 20 years of pharmaceutical industry experience, including the last 17 years focused solely on vaccines. He began his career at American Home Products in Clinical Research in 1984. He joined Aventis Pasteur (then Connaught Labs) in 1987 as clinical research coordinator and has held research and development positions of increasing responsibility, including clinical research manager and director of clinical operations. Mr. Hosbach also served as project manager for the development and licensure of Tripedia, the first diphtheria, tetanus, and acellular pertussis (DTaP) vaccine approved by FDA for use in U.S. infants. During his clinical research career at Aventis Pasteur, he contributed to the development and licensure of seven vaccines and has authored or coauthored several clinical research articles. From 2000 through 2002, Mr. Hosbach served on the board of directors for Pocono Medical Center, in East Stroudsburg, Pennsylvania. Since 2003 he has served on the board of directors of Pocono Health Systems, which includes Pocono Medical Center.

James M. Hughes, M.D., received his B.A. in 1966 and M.D. in 1971 from Stanford University. He completed a residency in internal medicine at the University of Washington and a fellowship in infectious diseases at the University of Virginia. He is board certified in internal medicine, infectious diseases, and preventive medicine. He first joined CDC as an epidemic intelligence service officer in 1973. During his CDC career, he has worked primarily in the areas of foodborne disease and infection control in health care settings. He became director of the NCID in 1992. The center is currently working to address domestic and global challenges posed by emerging infectious diseases and the threat of bioterrorism. He is a member of the IOM and a fellow of the American College of Physicians, the Infectious Diseases Society of America, and the AAAS. He is an Assistant Surgeon General in the Public Health Service.

Stephen A. Johnston, Ph.D., is currently director of the Center for Innovations in Medicine in the Biodesign Institute at Arizona State University (www.biodesign.asu.edu). His center focuses on formulating and implementing disruptive technologies for basic problems in healthcare. Currently the Center has three divisions:

Genomes to Vaccines, Cancer Eradication, and Doc-in-a-Box. The Genomes to Vaccines group has developed high-throughput systems to screen for vaccine candidates and is applying them to predict and produce chemical vaccines. The Cancer Eradication group is working on formulating a universal prophylactic vaccine for cancer. The Doc-in-a-Box group is developing technologies to facilitate pre-symptomatic diagnosis. Johnston founded the Center for Biomedical Inventions (a.k.a. Center for Translation Research) at the University of Texas-Southwestern, the first center of its kind in the medical arena. He and his colleagues have developed numerous inventions and innovations including the gene gun, genetic immunization, TEV protease system, organelle transformation, digital optical chemistry arrays, expression library immunization, linear expression elements, and others. He also was involved in transcription research for years, first cloning Gal4, then later discovering functional domains in transcription factors and the connection of the proteasome to transcription. He has been professor at the University of Texas Southwestern Medical Center at Dallas and associate and assistant professor at Duke University. He has been involved in several capacities as an advisor on biosecurity since 1996 and is a member of the WRCE SAB and a founding member of BioChem 20/20.

Gerald T. Keusch, M.D., is provost and dean for Global Health at Boston University and Boston University School of Public Health. He is a graduate of Columbia College (1958) and Harvard Medical School (1963). After completing a residency in internal medicine, fellowship training in infectious diseases, and two years as a NIH research associate at the SEATO Medical Research Laboratory in Bangkok, Thailand, Dr. Keusch joined the faculty of Mt. Sinai School of Medicine in 1970, where he established a laboratory to study the pathogenesis of bacillary dysentery and the biology and biochemistry of Shiga toxin. In 1979 he moved to Tufts Medical School and New England Medical Center in Boston to found the Division of Geographic Medicine, which focused on the molecular and cellular biology of tropical infectious disease. In 1986 he integrated the clinical infectious diseases program into the Division of Geographic Medicine and Infectious Diseases, continuing as division chief until 1998. He has worked in the laboratory and in the field in Latin America, Africa, and Asia on basic and clinical infectious diseases and HIV/AIDS research. From 1998 to 2003, he was associate director for international research and director of the Fogarty International Center at the NIH. Dr. Keusch is a member of the American Society for Clinical Investigation, the Association of American Physicians, the ASM, and the Infectious Diseases Society of America. He has received the Squibb (1981), Finland (1997), and Bristol (2002) awards of the Infectious Diseases Society of America. In 2002 he was elected to the IOM.

Rima F. Khabbaz, M.D., is director of the NCID at the CDC. She received her B.S. in 1975 and her M.D. in 1979 from the American University of Beirut in

Lebanon. She trained in internal medicine and completed a fellowship in infectious diseases at the University of Maryland in Baltimore. She is board certified in internal medicine. She first joined CDC as an epidemic intelligence service officer in 1980. During her CDC career she worked primarily in the areas of health care-associated infections and viral diseases. She is a fellow of the Infectious Diseases Society of America and an elected member of the American Epidemiologic Society. She served on the Blood Product Advisory Committee of the FDA, on the FDA's Transmissible Spongiform Encephalopathy Advisory Committee, and on the Annual Meeting Scientific Program Committee of the Infectious Diseases Society of America. She played a leading role in developing CDC's programs related to blood safety and food safety and in CDC's responses to outbreaks of new and reemerging diseases.

Lonnie J. King, D.V.M., is currently the director of CDC's new National Center for Zoonotic, Vector-Borne, and Enteric Diseases (NCZVED). In this new position Dr. King leads the center's activities for surveillance, diagnostics, disease investigations, epidemiology, research, public education, policy development and diseases prevention and control programs. NCZVED also focuses on water-borne, food-borne, vector-borne, and zoonotic diseases of public health concern, which also includes most of CDC's select and bioterrorism agents, neglected tropical diseases, and emerging zoonoses. Before serving as director, he was the first chief of the agency's Office of Strategy and Innovation. Dr. King was appointed dean of the College of Veterinary Medicine, Michigan State University, effective July 1, 1996, and became the college's eleventh dean since it was established by the Michigan legislature in 1910. He served for 10 years as dean of the college. As dean, he was the chief executive officer for academic programs, research, the teaching hospital, diagnostic center for population and animal health, basic and clinical science departments, and the outreach and continuing education programs. As dean and professor of large animal clinical sciences, Dr. King was instrumental in obtaining funds for the construction of the $60 million Diagnostic Center for Population and Animal Health, initiated the Center for Emerging Infectious Diseases in the college, served as the campus leader in food safety and had oversight for the National Food Safety and Toxicology Center. He brought the Center for Integrative Toxicology to the college and was the university's designated leader for counter-bioterrorism activities for his college and was involved in reestablishing public health programs at Michigan State University. Prior to this, Dr. King was administrator for the Animal and Plant Health Inspection Service (APHIS), U.S. Department of Agriculture, in Washington, D.C. In this role, he provided executive leadership and direction for ensuring the health and care of animals and plants, to improve agricultural productivity and competitiveness, and to contribute to the national economy and public health. He had served as administrator of APHIS since October 1992, and prior to that time was associate administrator. Dr. King served as the country's chief veterinary officer for 5 years and worked extensively in global trade agreements within NAFTA and the World Trade Organization. Before beginning his government career in 1977, Dr. King was in private veterinary practice for seven years in Dayton, Ohio, and Atlanta, Georgia. Prior to his current appointment, his assignments included field veterinary medical officer in Georgia and station epidemiologist in Texas. He spent five years in Hyattsville, Maryland, in staff assignments in Emergency Programs as well as Animal Health Information. While in Hyattsville, Dr. King directed the development of the agency's National Animal Health Monitoring System. He left APHIS briefly to serve as the Director of the Governmental Relations Division of the American Veterinary Medical Association (AVMA) in Washington, DC, and served as the lobbyist for the AVMA on Capitol Hill. From 1988-1991 Dr. King was the deputy administrator for Veterinary Services. In that position he was responsible for directing national veterinary and animal health programs, including the National Veterinary Services Lab and Plum Island Animal Disease Center. As a native of Wooster, Ohio, Dr. King received his Bachelor of Science and doctor of veterinary medicine degrees from the Ohio State University in 1966 and 1970, respectively. He earned his master of science degree in epidemiology from the University of Minnesota while on special assignment with the U.S. Department of Agriculture in 1980. He also received his master's degree in public administration from American University in Washington, DC in 1991. Dr. King has a broad knowledge of animal agriculture and the veterinary profession through his work with other governmental agencies, universities, major livestock and poultry groups, and private practitioners. Dr. King is a board-certified member of the American College of Veterinary Preventive Medicine and has completed the senior executive fellowship program at Harvard University. He served as president of the Association of American Veterinary Medical Colleges from 1999 to 2000 and was the vice-chair for the National Commission on Veterinary Economic Issues from 2000 to 2004. Dr. King helped start the National Alliance for Food Safety, served on the Governor's Task Force on Chronic Wasting Disease for the State of Michigan, and was a member of four NAS committees; most recently he chaired the National Academies Committee on Assessing the Nation's Framework for Addressing Animal Diseases. Dr. King is one of the developers of the Science, Politics, and Animal Health Policy Fellowship Program, and he lectures extensively on the future of animal health and veterinary medicine. He served as a consultant and member of the Board of Scientific Counselors to the CDC's National Center for Infectious Diseases, is a member of the IOM's Committee on Microbial Threats to Health, editor for the OIE Scientific Review on Emerging Zoonoses, is a current member of FDA's Board of Scientific Advisors, and is president of the American Veterinary Epidemiology Society. Dr. King was elected a member of the IOM in 2004.

Col. George W. Korch, Ph.D., is commander, United States Army Medical Research Institute for Infectious Diseases, Ft. Detrick, Maryland. Dr. Korch attended Boston University and earned a B.S. in Biology in 1974, followed by postgraduate study in mammalian ecology at the University of Kansas from 1975 to 1978. He earned his Ph.D. from the Johns Hopkins School of Hygiene and Public Health in Immunology and Infectious Diseases in 1985, followed by postdoctoral experience at Johns Hopkins from 1985 to 1986. His areas of training and specialty are the epidemiology of zoonotic viral pathogens and medical entomology. For the past 15 years he has also been engaged in research and program management for medical defense against biological pathogens used in terrorism or warfare.

Joshua Lederberg, Ph.D., is professor emeritus of molecular genetics and informatics and Sackler Foundation Scholar at the Rockefeller University in New York City. His lifelong research, for which he received the Nobel Prize in 1958, has been in genetic structure and function in microorganisms. He has a keen interest in international health and from 1990 to 1992 was co-chair of a previous IOM Committee on Emerging Microbial Threats to Health. Currently he is cochair of the Committee on Emerging Microbial Threats to Health in the Twenty-First Century. He has been a member of the NAS since 1957 and is a charter member of the IOM.

Lynn Marks, M.D., is board certified in internal medicine and infectious diseases. He was on faculty at the University of South Alabama College of Medicine in the Infectious Diseases Department focusing on patient care, teaching, and research, where his academic research interest was on the molecular genetics of bacterial pathogenicity. He subsequently joined anti-infectives clinical group of SmithKline Beecham (now GlaxoSmithKline) and later advanced to be global head of the Consumer Healthcare Division Medical and Regulatory Group. He then returned to pharmaceutical research and development as global head of the Infectious Diseases Therapeutic Area Strategy Team for GlaxoSmithKline.

Edward McSweegan, Ph.D., is a program officer at NIAID. He graduated from Boston College with a B.S. in 1978 and has an M.S. in microbiology from the University of New Hampshire and a Ph.D. in microbiology from the University of Rhode Island. He was a National Research Council Associate from 1984 to 1986 and did postdoctoral research at the Naval Medical Research Institute in Bethesda, Maryland. Dr. McSweegan served as a AAAS diplomacy fellow in the U.S. State Department from 1986 to 1988 and negotiated science and technology agreements with Poland, Hungary, and the former Soviet Union. After moving to the NIH, he continued to work on international health and science projects in Egypt, Israel, India, and Russia. Currently, he manages NIAID's bilateral program with India, the Indo–U.S. Vaccine Action Program, and represents NIAID in the HHS Biotechnology Engagement Program (BTEP) with Russia and related countries. He is a member of the AAAS, the ASM, and the DC

Science Writers Association. He is the author of numerous journal articles and science articles.

Stephen S. Morse, Ph.D., is founding director of the Center for Public Health Preparedness at the Mailman School of Public Health of Columbia University and is an associate professor in the epidemiology department. He recently returned to Columbia from four years in government service as program manager at the Defense Advanced Research Projects Agency (DARPA), where he co-directed the Pathogen Countermeasures Program and subsequently directed the Advanced Diagnostics Program. Before coming to Columbia, he was assistant professor of virology at the Rockefeller University in New York, where he remains an adjunct faculty member. He is the editor of two books, *Emerging Viruses* (Oxford University Press, 1993; paperback, 1996, which was selected by *American Scientist* for its list of 100 Top Science Books of the 20th Century, and *The Evolutionary Biology of Viruses* (Raven Press, 1994). He currently serves as a section editor of the CDC journal *Emerging Infectious Diseases* and was formerly an editor-in-chief of the Pasteur Institute's journal *Research in Virology*. Dr. Morse was chair and principal organizer of the 1989 NIAID/NIH Conference on Emerging Viruses, for which he originated the term and concept of *emerging viruses/infections*; has served as a member of the IOM-NAS Committee on Emerging Microbial Threats to Health, chaired its Task Force on Viruses, and was a contributor to the resulting report, *Emerging Infections* (1992); he was a member of the IOM's Committee on Xenograft Transplantation; he currently serves on the Steering Committee of the IOM's Forum on Emerging Infections (now the Forum on Microbial Threats); and he has served as an adviser to WHO, the Pan-American Health Organization, the FDA, the Defense Threat Reduction Agency, and other agencies. He is a fellow of the New York Academy of Sciences and a past chair of its microbiology section, a Fellow of the American Academy of Microbiology of the American College of Epidemiology, and an elected life member of the Council on Foreign Relations. He was the founding chair of ProMED, the nonprofit international Program to Monitor Emerging Diseases, and was one of the originators of ProMED-mail, an international network inaugurated by ProMED in 1994 for outbreak reporting and disease monitoring using the Internet. Dr. Morse received his Ph.D. from the University of Wisconsin-Madison.

Michael T. Osterholm, Ph.D., M.P.H., is director of the Center for Infectious Disease Research and Policy at the University of Minnesota, where he is also professor at the School of Public Health. Previously, Dr. Osterholm was the state epidemiologist and chief of the acute disease epidemiology section for the Minnesota Department of Health. He has received numerous research awards from the NIAID and the CDC. He served as principal investigator for the CDC-sponsored Emerging Infections Program in Minnesota. He has published more than 240 articles and abstracts on various emerging infectious disease problems and is the author of the best selling book, *Living Terrors: What America Needs to Know to Survive the Coming Bioterrorist Catastrophe*. He is past president of the Council of State and Territorial Epidemiologists. He currently serves on the NAS-IOM Forum on Emerging Infections. He has also served on the IOM Committee to Ensure Safe Food from Production to Consumption, the IOM Committee on the Department of Defense Persian Gulf Syndrome Comprehensive Clinical Evaluation Program, and as a reviewer for the IOM report on chemical and biological terrorism.

George Poste, Ph.D., D.V.M., is director of the Arizona Biodesign Institute and Dell E. Webb Distinguished Professor of Biology at Arizona State University. From 1992 to 1999, he was chief science and technology officer and president, Research and Development of SmithKline Beecham (SB). During his tenure at SB, he was associated with the successful registration of 29 drug, vaccine, and diagnostic products. He is chairman of diaDexus and Structural GenomiX in California and Orchid Biosciences in Princeton. He serves on the board of directors of AdvancePCS and Monsanto. He is an advisor on biotechnology to several venture capital funds and investment banks. In May 2003, he was appointed as director of the Arizona Biodesign Institute at Arizona State University. This is a major new initiative combining research groups in biotechnology, nanotechnology, materials science, advanced computing, and neuromorphic engineering. He is a fellow of Pembroke College at Cambridge and distinguished fellow at the Hoover Institution and Stanford University. He is a member of the Defense Science Board of the U.S. Department of Defense and in this capacity he chairs the Task Force on Bioterrorism. He is also a member of the NAS Working Group on Defense Against Bioweapons. Dr. Poste is a board certified pathologist, a fellow of the Royal Society, and a fellow of the Academy of Medical Sciences. He was awarded the rank of Commander of the British Empire by Queen Elizabeth II in 1999 for services to medicine and for the advancement of biotechnology. He has published over 350 scientific papers, has coedited 15 books on cancer, biotechnology, and infectious diseases, and serves on the editorial board of multiple technical journals. He is routinely invited to be the keynote speaker at a wide variety of academic, corporate, investment, and government meetings to discuss the impact of biotechnology and genetics on health care and the challenges posed by bioterrorism.

David A. Relman, M.D., Ph.D., is an associate professor of medicine (infectious diseases and geographic medicine) and of microbiology and immunology at Stanford University School of Medicine in Stanford, California, and chief of the infectious disease section at the Veterans Affairs (VA) Palo Alto Health Care System in Palo Alto, California. Dr. Relman received his B.S. in biology from the Massachusetts Institute of Technology, in Cambridge and his M.D. from Harvard Medical School. He completed his residency in internal medicine and a clinical fellowship in infectious diseases at Massachusetts General Hospital, Boston, after which he moved to Stanford in 1994. His major focus is laboratory research directed toward characterizing the human endogenous microbial flora, host-microbe interactions, and identifying previously unrecognized microbial pathogens using molecular and genomic approaches. He has described a number of new human microbial pathogens. Dr. Relman's lab (http://relman.stanford.edu) is currently exploring human oral and intestinal microbial ecology, sources of variation in host genome-wide expression during responses to infection and also during states of health, and how *Bordetella* species, including the agent of whooping cough, cause disease. He has published over 150 peer-reviewed articles, reviews, editorials, and book chapters on pathogen discovery and bacterial pathogenesis. He has served on scientific program committees for the ASM and the Infectious Diseases Society of America (IDSA) and on advisory panels for NIH, CDC, the departments of energy and defense, and the National Aeronautics and Space Administration. He was co-chair of the Committee on Advances in Technology and the Prevention of their Application to Next Generation Biowarefare Threats for the NAS. He is a member of the board of directors of the IDSA and the Board of Scientific Counselors at the National Institute of Dental and Craniofacial Research at the NIH. He received the Squibb Award from IDSA in 2001 and the Senior Scholar Award in Global Infectious Diseases from the Ellison Medical Foundation in 2002, and he is a fellow of the American Academy of Microbiology.

Gary A. Roselle, M.D., received his M.D. from the Ohio State University School of Medicine in 1973. He served his residency at the Northwestern University School of Medicine and his infectious diseases fellowship at the University of Cincinnati School of Medicine. He is the program director for infectious diseases for the VA Central Office in Washington, D.C., as well as the chief of the medical service at the Cincinnati VA Medical Center. He is a professor of medicine in the Department of Internal Medicine, Division of Infectious Diseases at the University of Cincinnati College of Medicine. Dr. Roselle serves on several national advisory committees. In addition, he is currently heading the Emerging Pathogens Initiative for the Department of Veterans Affairs. He has received commendations from the Cincinnati Medical Center director, the under secretary for health for the Department of Veterans Affairs, and the secretary of veterans affairs for his work in the infectious diseases program for the Department of Veterans Affairs. He has been an invited speaker at several national and international meetings and has published over 80 papers and several book chapters.

Janet Shoemaker is director of the ASM's Public Affairs Office, a position she has held since 1989. She is responsible for managing the legislative and regulatory affairs of this 42,000-member organization, the largest single biological science society in the world. She has served as principal investigator for a project funded by the National Science Foundation (NSF) to collect and disseminate data on the job market for recent doctorates in microbiology and has played a key role in ASM projects, including the production of the ASM *Employment Outlook in the Microbiological Sciences* and *The Impact of Managed Care and Health System Change on Clinical Microbiology*. Previously, she held positions as assistant director of public affairs for ASM, as ASM coordinator of the U.S./U.S.S.R. Exchange Program in Microbiology, a program sponsored and coordinated by the NSF and the U.S. Department of State, and as a freelance editor and writer. She received her baccalaureate, cum laude, from the University of Massachusetts, and is a graduate of the George Washington University programs in public policy and in editing and publications. She has served as commissioner to the Commission on Professionals in Science and Technology, and as the ASM representative to the ad hoc Group for Medical Research Funding, and is a member of Women in Government Relations, the American Society of Association Executives, and the AAAS. She has coauthored published articles on research funding, biotechnology, biological weapons control, and public policy issues related to microbiology.

Brian Staskawicz, Ph.D., is professor and chair, Department of Plant and Microbial Biology, University of California, Berkeley. Dr. Staskawicz received his B.A. in Biology from Bates College in 1974 and his Ph.D. from the University of California, Berkeley in 1980. Dr. Staskawicz's work has greatly contributed to understanding the molecular interactions between plants and their pathogens. He was elected to the NAS in 1998 for elucidating the mechanisms of disease resistance, as his lab was the first to clone a bacterial effector gene from a pathogen and among the first to clone and characterize plant disease-resistance genes. Dr. Staskawicz's research focuses on the interaction of the bacteria, Pseudomonas and Xanthomonas, with Arabidopsis, tomato and pepper. He has published extensively in this area and is a one of the leading scientists in the world working on elucidating the molecular basis of plant innate immunity.

Terence Taylor is president and director of the International Council for the Life Sciences (ICLS). He is responsible for the overall direction of the ICLS and its programs, which have the goal of enhancing global biosafety and biosecurity. Previously he was assistant director of the International Institute for Strategic Studies (IISS) (1995 to 2005), a leading independent international institute and president and executive director of its US office (2001 to 2005). He studies international security policy, risk analysis, scientific and technological developments and their impact on political and economic stability worldwide. At IISS he was one of the Institute's leading experts on issues associated with nuclear, biological, and chemical weapons and their means of delivery. In his previous appointments he has had particular responsibilities for issues affecting public safety and security in relation to biological risks and advances in the life sciences. He was one of the Commissioners to the UN Special Commission on Iraq for which he also conducted missions as a Chief Inspector. He was a Research Fellow on the Science Program at the Center for International Security and Cooperation at Stanford University where he carried out, among other subjects, studies of the implications for government and industry of the weapons of mass destruction treaties and agreements. He has also carried out consultancy work for the International Committee of the Red Cross on the implementation and development of the laws of armed conflict. He has served as Chairman of the World Federation of Scientists' Permanent Monitoring Panel on Risk Analysis. He served as a career office in the British Army on operations in many parts of the world including counterterrorist operations and UN peacekeeping. His publications include monographs, book chapters and articles for, among others, Stanford University, the World Economic Forum, Stockholm International Peace Research Institute (SIPRI), the Crimes of War Project, *International Herald Tribune, Wall Street Journal*, the *International Defence Review*, the *Independent* (London), *Tiempo* (Madrid), the *International and Comparative Law Quarterly*, the *Washington Quarterly* and other scholarly journals including unsigned contributions to IISS publications.